# California Science

## Science Content Support

## Grade 5

Harcourt
SCHOOL PUBLISHERS

Visit *The Learning Site!*
www.harcourtschool.com

Printed in the United States of America

ISBN-13: 978-0-15-352284-0
ISBN-10: 0-15-352284-4

2 3 4 5 6 7 8 9 10 022 17 16 15 14 13 12 11 10 09 08

# Contents



## Getting Ready for Science



## Unit 1 • Elements and Compounds

## Unit 2 • Structures of Living Things

## Unit 3 • The Water Cycle



## Unit 4 • Weather

## Unit 5 • The Solar System



## Vocabulary

Name ______________________________

Date ______________________________

## Getting to Know Your Textbook

**Welcome to *Science* by Harcourt School Publishers. You can look forward to an exciting year of discovery.**

**Your textbook has many features that can help you learn science this year. Use this scavenger hunt to learn more about it.**

1. What animal is on the cover of your book? ______________________

   Name one fact about the animal. ______________________

2. Find the unit called "Getting Ready for Science." Name one of the science tools you will use this year. ______________________

3. Name two of the steps of the Scientific Method, found in the same unit.

   ______________________

4. What are the three handbooks in the back of your book?

   ______________________

5. The Big Idea is what you will understand by the end of the unit. What's the "Big Idea" for Unit 5? ______________________

6. In order to understand the Big Idea, you will be answering Essential Questions. The Essential Questions are also the titles of the lessons. What is one of the Essential Questions for Unit 3?

   ______________________

7. What Reading Focus Skill is used in Lesson 2 of Unit 2?

8. What is the first word in the glossary? ______________________

9. What is the last word listed in the index? ______________________

10. Find an activity that you would like to try on one of the Make Connections pages. Write its title and page number. ______________________

Name ______________________

**11.** List the numbers of the California Standards that you will learn about in Unit 2, Lesson 1. ______________________

**12.** Name the title and page number of an Investigate you would like to try. What's the Investigation Skill Tip? ______________________

**13.** Name the title and page number of an Insta-Lab you would like to try.

______________________

**14.** Find a California on Location feature that you find interesting. What's the location?

______________________

**15.** Find the postcard at the beginning of Unit 3. Who's it from? Where was it sent from?

______________________

**16.** You can use the Vocabulary Preview to learn new science words from each lesson. The preview shows you how to say, or pronounce, the term. Find the Vocabulary Preview for Unit 2, Lesson 1. List two vocabulary words that you will learn. What photos are shown to help you understand those words?

______________________

**17.** The words in the Vocabulary Preview also appear in the text of your book. They are highlighted in yellow and are used in a way that helps to explain their meanings. Find the two words from the item above in the text and list the page number each appears on. ______________________

**18.** Find a California Fast Fact that you find interesting. Write its title and page number.

______________________

**19.** Write the name of a person featured in one of the People features.

______________________

**20.** Write three new things you expect to learn about this year.

______________________

______________________

______________________

Name ____________________

Date ____________________



# Lesson 1—What Tools Do Scientists Use?

## A. Explore Word Meanings

**Match the clue on the left with the term on the right.**

| Clue | Term |
|---|---|
| ___ the amount of matter in an object | **a.** forceps |
| ___ a measure of gravity's pull | **b.** balance |
| ___ a tool used to pick up or hold small objects | **c.** mass |
| ___ a tool used to measure the length and width of objects | **d.** weight |
| ___ a tool used to measure forces, such as weight or friction | **e.** microscope |
| ___ a tool used to measure the volume of a liquid | **f.** ruler |
| ___ a tool that makes small objects appear larger | **g.** spring scale |
| ___ how much space something takes up | **h.** graduated cylinder |
| ___ a tool that measures the amount of matter in an object | **i.** volume |

## B. Suffixes

The suffix –ative means "of, relating to." Use this information to draw a line between each word and its definition.

|  |  |
|---|---|
| qualitative observation | observations relating to the quantity, or amount, of something |
| quantitative observation | observations relating to the quality or kind of something |

Name ____________________

Date ____________________



# Lesson 1—What Tools Do Scientists Use?

## Connect Ideas

**Graphic organizers are drawings that help you organize information. They can help you connect ideas.**

- One kind of organizer is a bubble map. On a bubble map, the main idea is written in the center bubble. The ideas that are related to the main idea are written in the surrounding bubbles.

**Complete the bubble map with the tools you would use to make observations.**

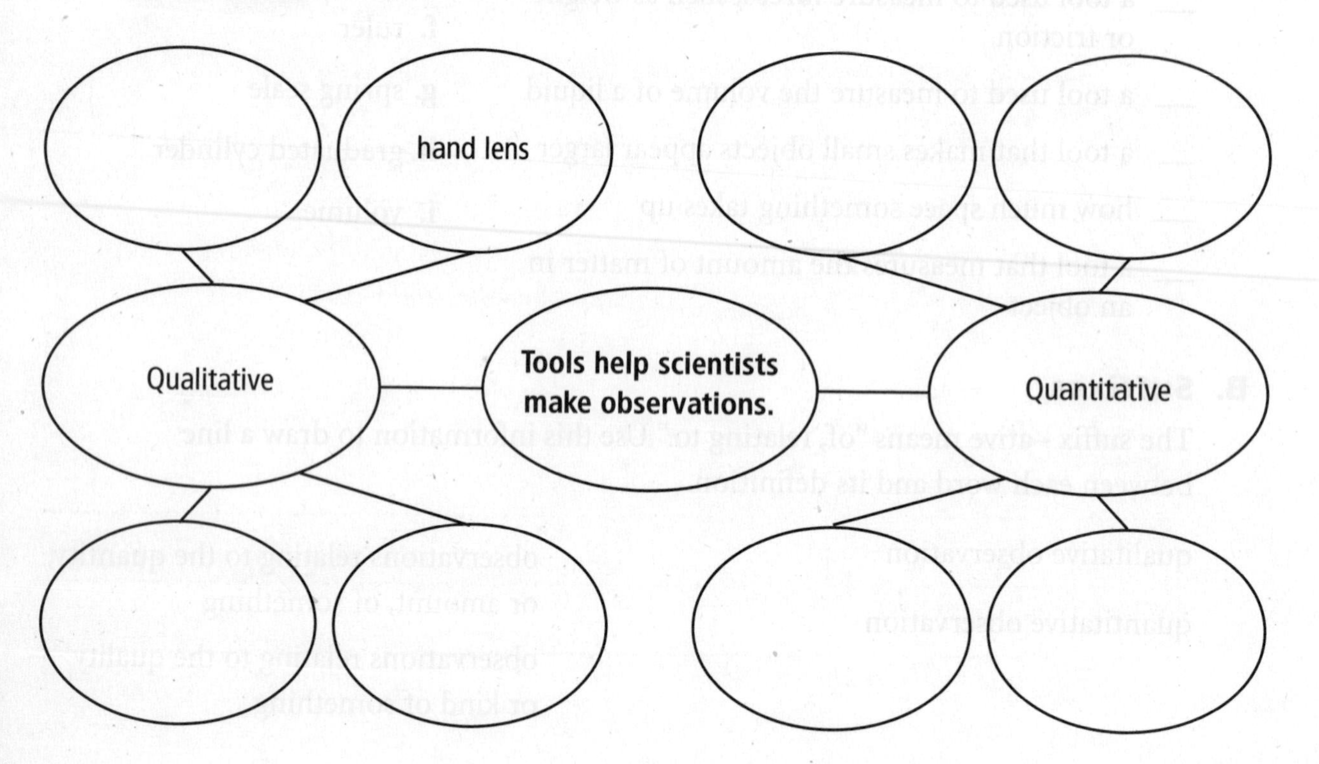

Name ______________________________

Date ______________________________



# Lesson 1—What Tools Do Scientists Use?

## 1. Investigation Skill Practice–Measure

**Draw a line to connect each qualitative observation with the necessary tool.**

| | |
|---|---|
| the length and width of a book | graduated cylinder |
| the mass of your shoe | spring scale |
| the weight of your pet dog | pan balance |
| the volume of water in a container | ruler |

## 2. Focus Skill Reading Focus Skill Practice–Main Idea and Details

**Read the selection. Underline the main idea. List at least two details about the main idea.**

Do you wear glasses? Did you know that people have been wearing glasses for hundreds of years? Eyeglasses, or spectacles, have been around since the thirteenth century. The first pair of wearable glasses that we know of was invented in Italy by Salvino D'Armate. The year was 1284! In 1752, James Ayscough invented glasses that had green or blue lenses. He thought that the light from clear lenses was hard on the eyes. They were a little bit like our modern sunglasses!

______________________________________________________________

______________________________________________________________

______________________________________________________________

Name ______________________

## Science Concepts

**3. Circle the word or words in each sentence that make the sentence true.**

**a.** Scientists use microscopes to make (quantitative/qualitative) observations, which do not involve measurements or numbers.

**b.** Thermometers and spring scales are tools that help scientists make (quantitative/qualitative) observations.

**c.** A (microscope/thermometer) lets you see details that you could not see with your eyes alone.

**d.** A curved glass can (disappear/magnify) objects, or make them look larger.

**e.** Anton van Leeuwenhouk used a lens to see creatures in a drop of water. He called them (animalcules/molecules).

**f.** The main parts of a microscope include the (earpiece/eyepiece), stage, base, nosepiece, and adjustment knobs.

**g.** Galileio Galilei made the first (thermometer/microscope) in 1592.

**h.** Fahrenheit made a thermometer using (mercury/hydrogen) in a sealed glass tube and marked units called degrees on the tube.

**i.** The Celsius scale has (100/1000) degrees between the freezing point of water and the boiling point of water.

**4. Draw a line to connect tool on the left with the unit of measure on the right.**

| | |
|---|---|
| spring scale | grams or kilograms |
| balance | centimeters and millimeters |
| graduated cylinder | newtons |
| ruler | milliliters or liters |

Name ______________________________

Date ______________________________

# Lesson 2—What Inquiry Skills Do Scientists Use?

## A. Explore Word Meanings

**For each sentence, fill in the blank with a word or words from the box that best completes the sentence.**

| investigation | inquiry | experiment | dependent variable | control variable |
|---|---|---|---|---|

1. The thing that is being controlled or measured in an investigation is called the ________________

2. An organized way to gather information so you can answer questions is called an ________________

3. An ________________ is a procedure that is carried out to gather data about an object or an event.

4. A procedure you carry out under controlled conditions to test a hypothesis is an ________________

5. A ________________ is the part of an investigation that remains the same so the dependent variable can be measured.

## A. Explore Word Meanings

**Draw a line to match the term on the left with the clue on the right.**

| | |
|---|---|
| observe | use data to draw conclusions about things you haven't directly observed |
| predict | decide how two things are the same |
| compare | group or organize objects or events into categories |
| classify | use observations and data to form an idea of what will happen |
| infer | use your senses to gather information about objects and events |

Name ______________________

Date ______________________

Study Skills



# Lesson 2—What Inquiry Skills Do Scientists Use?

## Pose Questions

**Asking questions as you read can help you understand what you are learning.**

- Form questions as you read. Think about how ideas and events are related.
- Use the questions to guide your reading. Look for the answers as you read.

**Complete the chart with your own questions about what you read. Then, answer the questions with information you find in the lesson.**

| What Inquiry Skills Do Scientists Use? | |
|---|---|
| **Questions** | **Answers** |
| 1. How is inquiry different from asking questions about something? | Inquiry is an organized way of getting information. You can be curious about something without trying to get information to answer your questions. |
| 2. | |
| 3. | |
| 4. | |
| 5. | |

Name ________________________________

Date ________________________________

# Lesson 2—What Inquiry Skills Do Scientists Use?

1. **Investigation Skill Practice–Develop a Testable Question**

**Suppose you want to find out how sunlight affects rose bushes. You want to perform an experiment in which rose bushes are exposed to various amounts of sunlight. Which of the following questions is a testable question for this experiment?**

**1.** How do rose bushes use sunlight to grow?

**2.** Why do rose bushes grow better in more light?

**3.** How do different amounts of sunlight affect rose bushes?

2. Focus Skill **Reading Focus Skill Practice–Main Idea and Details**

**Read the selection. Underline the main idea. Write two details that support the main idea.**

Even though you may not notice, you are using inquiry skills all the time. When you walk in your house and smell food cooking, you use your senses to observe and draw conclusions about what is for dinner based on that information. When you put away clean clothes, you may classify them—sort them into groups of shirts, pants, and socks. You may compare one batter's swing with another batter's swing at a baseball game. How many other inquiry skills do you use every day?

______________________________________________________________

______________________________________________________________

______________________________________________________________

Name ______________________________

## Science Concepts

**3. Check (✓) the statements below that agree with the information found in the lesson.**

_____ You can use your senses to gather information about objects and events.

_____ You should never ask questions about an object or event.

_____ Inquiry skills are skills that help you answer questions about an object or event.

_____ If you study an object and diagrams of the object, you are observing the object.

_____ If you compare two objects, you group them into categories based on certain characteristics.

_____ When you infer, you use logical reasoning to come to a conclusion based on data and observations.

_____ The first step in an investigation is to develop a testable question.

_____ The control variable in an investigation is the variable being measured.

**4. Circle the word that best completes each sentence.**

A. You can use your ____________ from an experiment to draw conclusions.

Predictions    data    hypothesis

B. You can use a ____________ to test something before building the real thing.

model    inquiry    conclusion

C. You can use tables, reports, and diagrams to ____________ the results of your experiment to others.

predict    communicate    hypothesize

D. When you ____________, you suggest an outcome or explanation that can be tested in an experiment.

inquire    communicate    hypothesize

Name ______________________________

Date ______________________________



# Lesson 2—What Inquiry Skills Do Scientists Use?

## Identifying Variables

A scientist wanted to determine whether the color of a house's roof would affect how much heat the house would absorb. The scientist decided to perform an experiment to test this idea. She set up six identical birdhouses together in a field and put thermometers inside each birdhouse to record the temperature during the day. Three of the birdhouses had black roofs, and three of the birdhouses had white roofs. She recorded the temperatures of each birdhouse at 9 am, noon, and 3 pm each day for a week.

**1.** Which variable or variables in the experiment did the scientist change?

______________________________

______________________________

**2.** Which variable or variables in the experiment did the scientist keep the same?

______________________________

______________________________

______________________________

**3.** What was the dependent variable that was being measured in the experiment?

______________________________

**4.** Which question above describes the control variable?

______________________________

Name ______________________________

A scientist has developed a new type of fertilizer for lawns. The scientist hypothesizes that using the fertilizer will lead to lawns with greener, thicker grass. The scientist wants to run an experiment that will test whether the new fertilizer is more effective than a traditional lawn fertilizer.

**Use the information above to answer the questions below.**

**1.** In the experiment, name some variables that the scientist will want to keep the same.

______________________________

______________________________

______________________________

______________________________

**2.** Describe the variable that the scientist will change and how the variable will be changed in the experiment.

______________________________

______________________________

______________________________

**3.** Explain what things the scientist might use as a dependent variable, or the variable that would be measured as part of the experiment.

______________________________

______________________________

______________________________

**4.** Describe some results that would support the scientist's hypothesis.

______________________________

______________________________

______________________________

Name ______________________

Date ______________________

# Lesson 3—How Do Scientists Record and Interpret Data?

## A. Explore Word Meanings

**Draw a line to match the clue on the left to the term on the right.**

| | |
|---|---|
| a sketch or other visual representation that shows an idea or object | criteria |
| putting objects into groups based on your criteria | classify |
| specific qualities you use to put objects into groups | conclusion |
| a decision you make based on information | diagram |

## B. Graphs and Diagrams

**Complete the table with information you can illustrate using different types of graphs and diagrams.**

| Type of graph or diagram | Uses |
|---|---|
| bar graph | used to compare information about different objects, events, or groups |
| line graph | |
| circle graph | |
| diagram | |

Name ___________________

Date ___________________

**Study Skills**



# Lesson 3—How Do Scientists Record and Interpret Data?

## Organize Information

**A graphic organizer can help you make sense of the facts you read.**

- Tables, charts, and webs are graphic organizers that can show main ideas and important details.
- A graphic organizer can help you classify and categorize information. It can also help you understand the relationship between the subject of the chapter and each lesson.

**Complete the graphic organizer with details about ways that scientists record and interpret data.**

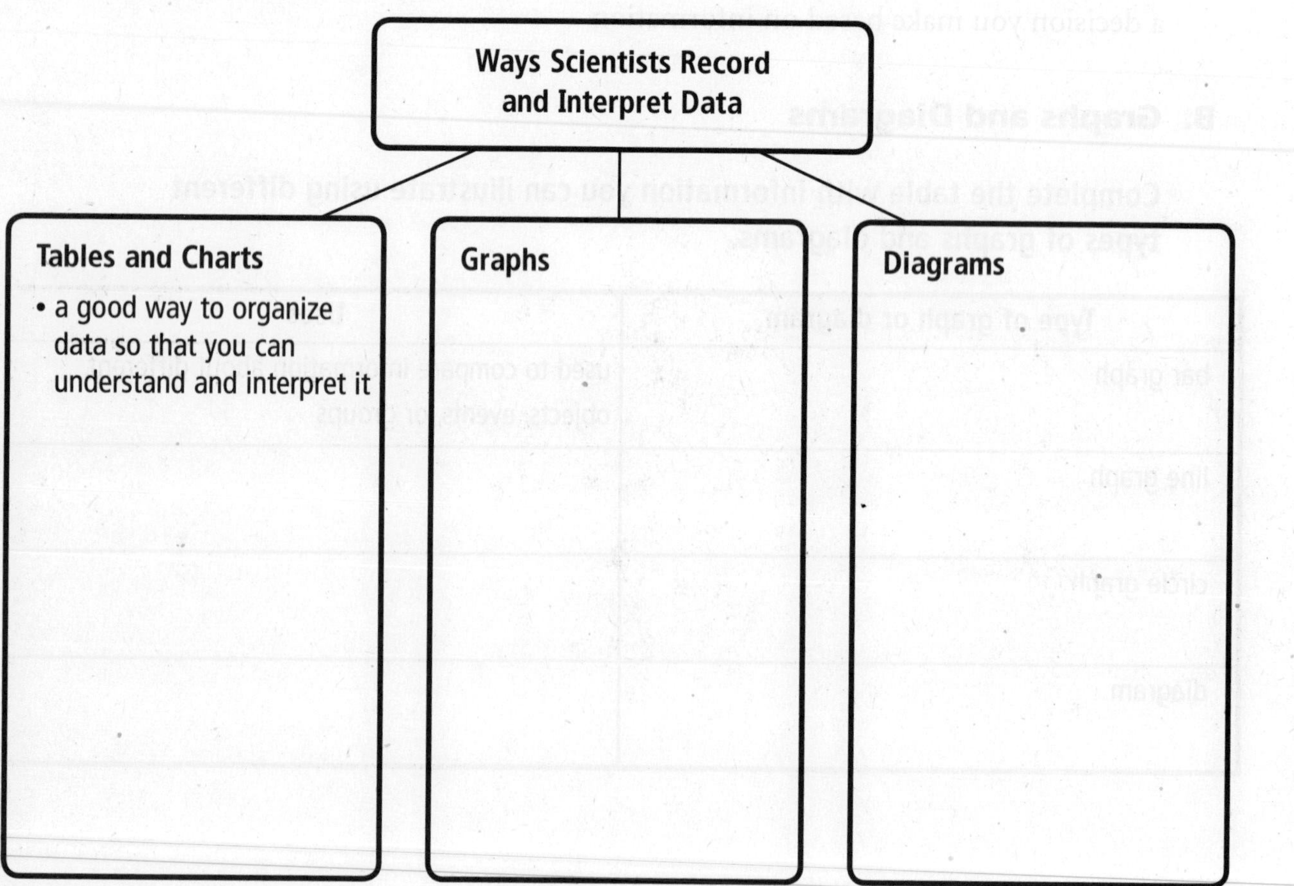

Name ________________________________

Date ________________________________

# Lesson 3—How Do Scientists Record and Interpret Data?

## 1. Investigation Skill Practice–Record Data

**Suzanne collected information about the temperature at noon in her town each day for a week. Sunday it was 24°C and sunny. Monday it was 23°C and sunny. Tuesday it rained and the temperature fell to 16°C. Wednesday it rained and was 19°C. The next three days were sunny and were 20°C, 23°C, and 23°C, in that order. Complete the table using the data Suzanne collected.**

| Day | Temperature | Rain or Sun? |
|---|---|---|
| Sunday | 24°C | sun |
| | 23°C | |
| | 16°C | |
| | 19°C | |
| | 20°C | |
| | 23°C | |
| | 23°C | |

## 2. Focus Skill Reading Skill Practice–Main Idea and Details

**Circle the details that support the main idea.**

Main Idea: Tables and charts can help you draw conclusions from your data.

1. Tables and charts are good ways to organize data.

2. Tables and charts make it harder to interpret data.

3. You can record information in a table as you collect it.

4. Tables and charts make it easier to see patterns in your data.

5. A bar graph can be used to display and compare data.

Name ______________________

## Science Concepts

**3. Decide which word completes each sentence. Write the letter in the blank.**

_____ When you __________ objects, you group them by specific qualities.

_____ If you group single-engine planes together and twin-engine planes together, you are using the number of engines as your __________.

_____ Tables and charts are both ways to __________ data so that you can understand and interpret it.

_____ As you __________ data during an investigation, you can write it in a table.

_____ You use the information you collect during an investigation to draw a __________.

**a.** criteria

**b.** classify

**c.** conclusion

**d.** organize

**e.** collect

**4. Use the words in the box to complete each sentence.**

| bar graph | diagram | patterns | whole |
|---|---|---|---|

**a.** Both a table and a __________ can be used to compare information about objects, groups, or events.

**b.** Looking for __________ in your data can help you draw conclusions.

**c.** A circle graph can show the parts of a __________.

**d.** You can use a __________ to show the inside parts of a submarine and how the parts work together.

Name ______________________

Date ______________________

# Lesson 3—How Do Scientists Record and Interpret Data?

## Recording/Interpreting Data

**A. Several scientists worked throughout a state to count the populations of four different types of birds at three different locations in the state. They organized the data in a table, which is shown below. Use the data table to answer the questions.**

| | Robin | Wren | Sparrow | Cardinal |
|---|---|---|---|---|
| Location #1 | 243 | 112 | 355 | 210 |
| Location #2 | 56 | 45 | 250 | 199 |
| Location #3 | 178 | 87 | 289 | 156 |

**1.** Create a row that shows the total number of each bird observed in the state.

**2.** Which bird was present in the greatest numbers?

______________________________________________

**3.** Which bird had the lowest population in the state?

______________________________________________

**4.** Create a graph that shows the total populations of each type of bird.

Name ____________________

**B. A student found the chart shown below. It shows some of the volcanoes that have erupted in western United States.**

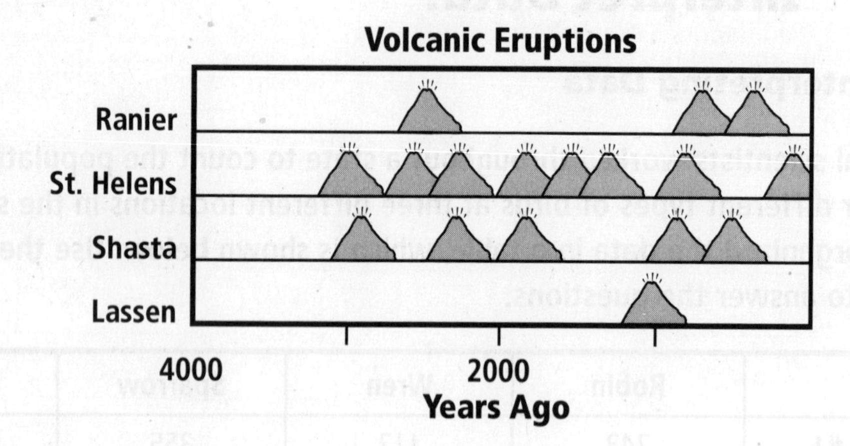


**1.** How long ago was the first volcanic eruption?

____________________

**2.** Which volcano has erupted the most times?

____________________

**3.** Which eruption was most recent?

____________________

**4.** Make a bar graph showing the number of eruptions of each of the volcanoes. Arrange the graph to show the number of eruptions from least to most.

Name ______________________

**C. The following chart shows the numbers of gas and oil wells in four counties of California. The data also shows the amount of gas and oil produced from each.**

| Country | Number of Oil Wells | Number of Gas Wells | Barrels of Oil produced | Gas produced (mcf) | Total Number of wells |
|---|---|---|---|---|---|
| Fresno | 1,966 | 6 | 7,218,816 | 2,575,047 | |
| Kern | 36,955 | 142 | 196,774,148 | 209,394,641 | |
| Los Angeles | 3,189 | 5 | 27,608,872 | 11,921,115 | |
| Ventura | 1,839 | 7 | 8,508,570 | 6,389,527 | |
| Totals | | | | | |

1. Fill in the column that shows the total number of wells by adding the number of oil wells plus the number of gas wells. Fill in the row to show the total numbers of wells and production. (You may use a calculator to help you.)

2. What can you conclude about the number of units of gas (mcf) produced per well as compared to the total number of units of oil (barrels) produced per well?

______________________

______________________

______________________

3. Show the number of oil wells found in these four counties in a pie graph. Color Fresno County in blue, Kern County in red, Los Angeles County in yellow, and Ventura County in green.

Name ______________________________

D. **There are eight National Parks in the state of California. The total number of acres and the number of acres owned by the federal government in each park is shown below. Use this data table to answer the questions.**

| Park Name | Total Acres | Federal Acres | |
|---|---|---|---|
| Channel Islands | 249,353 | 70,519 | |
| Death Valley | 3,367,628 | 3,348,929 | |
| Joshua Tree | 1,022,976 | 782,829 | |
| Kings Canyon | 461,901 | 461,845 | |
| Lassen Volcanic | 106,372 | 106,366 | |
| Redwood | 105,516 | 71,715 | |
| Sequoia | 402,510 | 402,334 | |
| Yosemite | 761,266 | 759,530 | |

1. Fill in the column at the right that shows the number of nonfederal acres in each park.
2. Which park has the smallest number of nonfederal acres?

______________________________

3. Make a bar graph that shows the parks from smallest to largest.

Name ______________________________

Date ______________________________

# Lesson 4—What Is the Scientific Method?

## A. Explore Word Meanings

**Match the clue on the left to the term on the right. Write the letter in the blank.**

| Clue | Term |
|---|---|
| _____ a series of steps used to plan and carry out investigations | a. report |
| _____ the information you gather during an investigation | b. evidence |
| _____ a possible answer to a question, which can be tested | c. scientific method |
| _____ a written account of the findings of your investigation | d. hypothesis |

## B. Sequence

**Use the numbers 1 through 5 to put the steps of the scientific method in the correct order.**

_____ Plan an investigation

_____ Draw conclusions and write a report

_____ Form a hypothesis

_____ Conduct the investigation

_____ Observe and ask questions

Name ______________________

Date ______________________

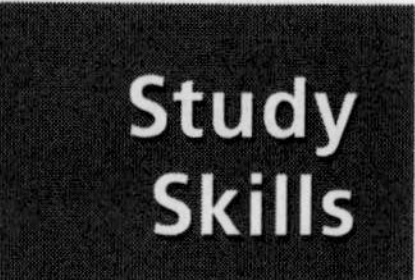




# Lesson 4—What Is the Scientific Method?

## Use an Anticipation Guide

**An anticipation guide can help you anticipate, or predict, what you will learn as you read.**

**Complete the anticipation guide using information from the lesson.**

| Observe, and Ask Questions | | |
|---|---|---|
| **Reading Check** | **Prediction** | **Correct?** |
| What can help you form a testable question? | your observations and other experiments | yes |
| **Form a Hypothesis** | | |
| **Reading Check** | **Prediction** | **Correct?** |
| ______ ______ | ______ ______ | |
| **Plan an Investigation** | | |
| **Reading Check** | **Prediction** | **Correct?** |
| ______ ______ | ______ ______ | |
| **Conduct an Investigation** | | |
| **Reading Check** | **Prediction** | **Correct?** |
| ______ ______ | ______ ______ | |
| **Draw Conclusions** | | |
| **Reading Check** | **Prediction** | **Correct?** |
| ______ ______ | ______ ______ | |
| **Write a Report** | | |
| **Reading Check** | **Prediction** | **Correct?** |
| ______ ______ | ______ ______ | |

Name ______________________________

Date ______________________________

# Lesson 4—What Is the Scientific Method?

1. **Investigation Skill Practice–Draw Conclusions**

**Suppose you fill two saucers with water. You set one outside on a warm, sunny day and one in a shady spot. After one day, the saucer in the sunny spot is empty. The saucer in the shade still has water in it. What conclusions can you draw from this information?**

______________________________________________________________

______________________________________________________________

______________________________________________________________

2. Focus Skill **Reading Skill Practice–Main Idea and Details**

**Read the selection. Underline the main idea. List at least two details.**

A good scientist always writes a report about his or her investigation, even when the results do not support the hypothesis. At first, the idea that you should tell everyone your hypothesis was wrong might seem strange. What do scientists do if the evidence from their experiment does not support their hypothesis? They write a report about the investigation. They can use details from the report to repeat the experiment. This way, they make sure the result wasn't just a mistake. Then they, or other scientists, can use the data from the investigation to form a new hypothesis and plan more investigations.

______________________________________________________________

______________________________________________________________

______________________________________________________________

Name ______________________________

## Science Concepts

**3. Circle the letter of the best answer to each question.**

1. What is the scientific method?

   A. A series of steps scientists use to plan and carry out investigations.

   B. Inquiry skills that you use in planning an experiment.

2. What do all investigations start with?

   A. an experiment

   B. a testable question

3. What is the second step in the scientific method?

   A. Perform an experiment.

   B. Form a hypothesis.

4. How could you test a hypothesis that a round-balloon rocket will travel a shorter distance than a long-balloon rocket?

   A. Conduct an investigation.

   B. Do research at the library.

5. What do you need to do before you conduct an investigation?

   A. Plan the investigation.

   B. Observe another person's investigation.

**4. Choose the word that completes each sentence.**

6. The results of an investigation will either (support/disprove) your hypothesis or fail to support it.

7. You must be able to (support/test) your hypothesis.

8. When you conduct an investigation, it is important to observe, measure, and (record/prove) everything that happens.

9. Identifying (patterns/problems) in your data can help you draw conclusions.

10. Writing a report about how you conducted your investigation and what the results were is a way to (observe/communicate).

Name ______________________________

Date ______________________________



# Lesson 4—What Is the Scientific Method?

## Applying Scientific Method

**A. A scientist decided to do an investigation on which trees lose their leaves earliest in the fall. Below is a list of steps in the investigation. Number the steps in the correct order so that the investigation is following the scientific method.**

**Step** _____ The scientist analyzed data. The data showed that the elm tree had lost its leaves by October 16th, the oak tree has lost its leaves by November 10th, and the maple tree had lost its leaves by November 25th.

**Step** _____ The scientist hypothesized that the maple tree would be the last tree to lose its leaves in the fall.

**Step** _____ The scientist observed all the trees in the area during the fall. He wrote down the day each type of tree had finally lost all its leaves.

**Step** _____ The scientist looked around at the trees in the area. He noticed that some trees seemed to lose their leaves early in the fall, while others still had them late in the season. He wondered if certain types of trees lost their leaves earlier than others.

**Step** _____ The scientist decided to observe all the trees in his area and record the dates in which different types of trees lost their leaves.

**Step** _____ The scientist wrote a report, describing the steps of the investigation, the data he collected, and his conclusion.

**B. Fill in the blanks below with some of the steps you would use to investigate whether using high-octane gasoline in a car would improve gas mileage. Make sure you describe the steps in the order they would occur.**

**Step 1** Your uncle drives his car to work every day for a year. During the year, he sometimes fills it with regular gasoline, and sometimes with premium-grade gasoline. It seems like it gets better gas mileage when he uses the premium-grade gasoline, but he isn't quite sure. You wonders if the premium-grade gasoline really does give better gas mileage.

Name ______________________

**Step 2** ______________________

**Step 3** ______________________

**Step 4** ______________________

**Step 5** You collect the data and use it to create an average gas mileage chart. The chart is shown below.

| | Mileage |
|---|---|
| Regular Gas | 22.5 |
| Premium Gas | 22.5 |

**Step 6** ______________________

Name ______________________

Date ______________________

# Lesson 1– What Are Atoms and Elements?

## A. Word Families

**Choose a word or words from the box that belongs to the same word family.**

| atom | element | periodic table | molecule | mixture |
|---|---|---|---|---|

1. atomic ______________
2. elementary ______________
3. periodically ______________
4. molecular ______________
5. mix ______________

## B. Context Clues

**Choose a word or words from the box to complete each sentence.**

| compound | periodic table | physical properties | molecule | element |
|---|---|---|---|---|

1. When a substance is made of two or more different elements, it is called a(n) ______________.
2. To help them remember the properties of each element, scientists refer to the ______________.
3. Color, shape, melting point, and freezing point are all ______________ of a substance.
4. Two or more atoms that are joined together are called a(n) ______________.
5. A substance made of only one kind of atom is a(n) ______________.

Name ______________________

Date ______________________

# Lesson 1– What Are Atoms and Elements?

## Use a K-W-L Chart

**A K-W-L chart can help you focus on what you already know about a topic and what you want to learn about it**

- Use the K column to list what you already know about atoms and elements.
- Use the W column to list what you want to learn about atoms and elements.
- Use the L column to list what you have learned about the topic from your reading.

**Complete the K-W-L chart as you read this lesson.**

| Atoms and Elements | | |
|---|---|---|
| **What I Know** | **What I Want to Learn** | **What I Learned** |
| • I know that an atom is the smallest part of a substance. | • Why is it called an atom? | • |
| • ______ | • ______ | • ______ |
| • ______ | • ______ | • ______ |
| • ______ | • ______ | • ______ |

Name ______________________

Date ______________________

# Lesson 1– What Are Atoms and Elements?

## 1. Investigation Skill Practice–Classify

**An element contains one kind of atom. A compound is made of atoms of two or more elements. Look at the list below. Classify each item as an element or a compound.**

| | | |
|---|---|---|
| hydrogen | salt | carbon |
| methane | nitrogen | propane |
| carbon dioxide | oxygen | water |

**Elements** **Compounds**

## 2. Focus Skill Reading Skill Practice–Main Idea and Details

**Read the selection. Underline the main idea. Write 2 details on the lines below.**

There are important pieces of information you can gather from the periodic table of the elements. You can tell if an element is metal or nonmetal. The metals are on the left. The nonmetals are on the right. You can also tell the element's symbol, name, atomic mass, and atomic number. These are useful for decoding the chemical name of a molecule.

______________________________________________

______________________________________________

______________________________________________

Name ______________________________

## Science Concepts

### 3. Read the Periodic Table

**Consider the compound for table salt: NaCl. Use the periodic table in your textbook to answer the following questions.**

What elements make up table salt?

______________________________

How many atoms of each element are in table salt?

______________________________

### 4. Physical Properties

**Use the words listed in the box to describe the physical properties of each substance. Then add your own descriptions of physical properties for each substance.**

| | | |
|---|---|---|
| **sticky** | **liquid at room temperature** | **able to dissolve in water** |
| **colorless** | **solid at room temperature** | **hard** |
| **odorless** | | |
| **lumpy** | | |

**A.** water

______________________________

______________________________

**B.** peanut butter

______________________________

______________________________

**C.** cardboard

______________________________

______________________________

**D.** window glass

______________________________

______________________________

Name ______________________________

Date ______________________________

# Lesson 1– What Are Atoms and Elements?

## The Periodic Table

### A. History of the Periodic Table

In the past, many chemists tried to organize the elements in a sensible way. They tried many different ways, but found best success when Dmitri Mendeleev created the Periodic Table of the Elements.

**Imagine that you are a chemist living in the early 1900s. You know the properties of the following 10 elements. Experiment with different ways to organize the elements into a chart. Discuss your ideas with others.**

| | | | | | |
|---|---|---|---|---|---|
| beryllium | solid | 4 protons | gray | combines easily with oxygen | high melting point |
| boron | solid | 5 protons | black or dark brown | combines with nitrogen | poor electrical conductor |
| carbon | solid | 6 protons | varies with form | combines with itself | comes in graphite, diamond, and amorphous forms |
| helium | gas | 2 protons | colorless | can combine with oxygen | low melting point |
| hydrogen | gas | 1 protons | colorless | combines with oxygen, fluorine, chlorine | most common element in universe |
| fluorine | gas | 9 protons | yellow | highly reactive with all elements | very corrosive |
| lithium | solid | 3 protons | silvery | combines with oxygen, nitrogen, carbon, and hydrogen | low density |
| neon | gas | 10 protons | varies | rarely combines with other elements, except fluorine | emits colors with electrical charge |
| nitrogen | gas | 7 protons | colorless | combines with boron, hydrogen, and itself | most abundant gas in atmosphere |
| oxygen | gas | 8 protons | colorless | combines with nearly all other elements including itself | poor conductor of heat and electricity |

Name ______________________________

## B. Using the Periodic Table

The Periodic Table we use today organizes the elements by atomic number. One reason it works so well is that the elements in each column have similar properties.

Periodic Table

Key: 11 — Atomic number; Na — Element symbol; Sodium — Element name

Metals; Metalloids (semimetals); Nonmetals

| | 1 | 2 | 3 | 4 | 5 | 6 | 7 | 8 | 9 | 10 | 11 | 12 | 13 | 14 | 15 | 16 | 17 | 18 |
|---|---|---|---|---|---|---|---|---|---|---|---|---|---|---|---|---|---|---|
| 1 | 1 H Hydrogen | | | | | | | | | | | | | | | | | 2 He Helium |
| 2 | 3 Li Lithium | 4 Be Beryllium | | | | | | | | | | | 5 B Boron | 6 C Carbon | 7 N Nitrogen | 8 O Oxygen | 9 F Fluorine | 10 Ne Neon |
| 3 | 11 Na Sodium | 12 Mg Magnesium | | | | | | | | | | | 13 Al Aluminum | 14 Si Silicon | 15 P Phosphorus | 16 S Sulfur | 17 Cl Chlorine | 18 Ar Argon |
| 4 | 19 K Potassium | 20 Ca Calcium | 21 Sc Scandium | 22 Ti Titanium | 23 V Vanadium | 24 Cr Chromium | 25 Mn Manganese | 26 Fe Iron | 27 Co Cobalt | 28 Ni Nickel | 29 Cu Copper | 30 Zn Zinc | 31 Ga Gallium | 32 Ge Germanium | 33 As Arsenic | 34 Se Selenium | 35 Br Bromine | 36 Kr Krypton |
| 5 | 37 Rb Rubidium | 38 Sr Strontium | 39 Y Yttrium | 40 Zr Zirconium | 41 Nb Niobium | 42 Mo Molybdenum | 43 Tc Technetium | 44 Ru Ruthenium | 45 Rh Rhodium | 46 Pd Palladium | 47 Ag Silver | 48 Cd Cadmium | 49 In Indium | 50 Sn Tin | 51 Sb Antimony | 52 Te Tellurium | 53 I Iodine | 54 Xe Xenon |
| 6 | 55 Cs Cesium | 56 Ba Barium | 57 La Lanthanum | 72 Hf Hafnium | 73 Ta Tantalum | 74 W Tungsten | 75 Re Rhenium | 76 Os Osmium | 77 Ir Iridium | 78 Pt Platinum | 79 Au Gold | 80 Hg Mercury | 81 Tl Thallium | 82 Pb Lead | 83 Bi Bismuth | 84 Po Polonium | 85 At Astatine | 86 Rn Radon |
| 7 | 87 Fr Francium | 88 Ra Radium | 89 Ac Actinium | 104 Rf Rutherfordium | 105 Db Dubnium | 106 Sg Seaborgium | 107 Bh Bohrium | 108 Hs Hassium | 109 Mt Meitnerium | | | | | | | | | |

| 58 Ce Cerium | 59 Pr Praseodymium | 60 Nd Neodymium | 61 Pm Promethium | 62 Sm Samarium | 63 Eu Europium | 64 Gd Gadolinium | 65 Tb Terbium | 66 Dy Dysprosium | 67 Ho Holmium | 68 Er Erbium | 69 Tm Thulium | 70 Yb Ytterbium | 71 Lu Lutetium |
|---|---|---|---|---|---|---|---|---|---|---|---|---|---|
| 90 Th Thorium | 91 Pa Protactinium | 92 U Uranium | 93 Np Neptunium | 94 Pu Plutonium | 95 Am Americium | 96 Cm Curium | 97 Bk Berkelium | 98 Cf Californium | 99 Es Einsteinium | 100 Fm Fermium | 101 Md Mendelevium | 102 No Nobelium | 103 Lr Lawrencium |

Each box tells the atomic number, name, and symbol of an element. The symbols come from the English, Latin, or other name for the element.

**1.** What is the atomic number for oxygen? ______________________

**2.** What does this number mean? ______________________________

## C. Learn some element symbols with the concentration game on the next page.

- Copy and cut out the cards or copy them onto index cards.
- Shuffle and place them face down in a grid.
- You and a partner take turns flipping over two cards. Keep the cards if you match an element name and symbol. If not, turn the cards over.
- Play until all the cards are matched. Shuffle and play again.

Name ______________________________

| | | | |
|---|---|---|---|
| H | Chlorine | Argon | O |
| Al | Helium | N | Oxygen |
| Nitrogen | Ar | Carbon | Na |
| He | Calcium | Cl | Aluminum |
| Sodium | C | Ca | Hydrogen |

Name ______________________________

**C. The Periodic Table shows patterns in the elements. Elements in the same column have similar properties. Answer the questions below.**

**1.** How does atomic number change across a row? Down a column?

______________________________

**2.** Where are the metals located in the Periodic Table? Where are the nonmetals and semimetals located?

______________________________

______________________________

**3.** Argon and xenon are gases. They do not combine easily with other elements. Identify another element with similar properties. Explain your answer.

______________________________

______________________________

**4.** Hydrogen and sodium combine easily with chlorine. Identify another element that combines easily with chlorine and explain your answer.

______________________________

______________________________

**5.** If you put lithium in water, not much happens. Add sodium to water and you will see a small amount of bubbling. Potassium and water make an explosive combination. Describe the pattern seen in column 1. Predict what might happen if you mix water and rubidium.

______________________________

______________________________

**6.** The size of the atom increases as you go down a column. What is the largest atom in column 16?

______________________________

**7.** The size of the atom decreases as you go across a row. What is the largest atom in row 3?

______________________________

Name ______________________________

Date ______________________________

# Lesson 2—What Are Metals?

## A. Graphic Organizer

**In the chart below, write the definition for each vocabulary word. Then give an example from the lesson that describes each word.**

| Vocabulary Word | Definition | Give an Example |
|---|---|---|
| metal | | |
| nonmetal | | |
| alloy | | |
| metalloid | | |
| malleable | | |

## B. Write sentences

**On the lines below, write a sentence for each of the words from the box.**

| malleable | metalloid | alloy |
|---|---|---|

______________________________

______________________________

______________________________

Name ________________________________

Date ________________________________

# Lesson 2—What Are Metals?

## Take Notes

**Taking notes can help you remember important ideas.**

- Write down important facts and ideas. Use your own words. You do not have to write in complete sentences.
- One way to organize notes is in a chart. Write down the main ideas in one column and facts and details in another.

**As you read this lesson, use the chart below to take notes.**

| What Are Metals? | |
|---|---|
| **Main Ideas** | **Fact** |
| • metals have high thermal conductivity—they conduct heat well | • a substance with high thermal conductivity could be used for a cooking pot |
| • ____________ | • ____________ |
| • ____________ | • ____________ |
| • ____________ | • ____________ |
| • ____________ | • ____________ |

Name ______________________

Date ______________________

Unit 1, Lesson 2

# Lesson 2—What Are Metals?

1. **Investigation Skill Practice–Infer**

**Caleb's dad is a jeweler. He makes necklaces and earrings out of silver. To do this, he pounds the silver into the shapes he wants. What can you infer about the physical properties of silver that make it good to use for making jewelry.**

______________________________________________

______________________________________________

**At Thanksgiving time, Anna's mother took out her good turkey platter. Anna noticed that it had brown stains on it. Her mother gave her a cloth and a special lotion to use to rub off the stains. What can you infer about the platter to explain why it got stains?**

______________________________________________

______________________________________________

______________________________________________

2. Focus Skill **Reading Skill Practice–Main Idea and Details**

**Read the selection. Underline the main idea. Write two details on the lines below.**

If you lift the aluminum foil off the top of a steaming hot casserole, you will not burn your finger. The heat leaves the foil as soon as it no longer touches the hot substance. If you cover a bowl with aluminum foil, you can wrap it very tightly and press the edges of the foil to form a tight seal around the top. These two physical properties make aluminum foil very useful in the kitchen.

______________________________________________

______________________________________________

______________________________________________

Name ___________________________

## Science Concepts

**3. For each metal in the top column, put an X in the row that describes one of its physical properties.**

| | germanium | silicon | bronze | steel | tin | iron | silver |
|---|---|---|---|---|---|---|---|
| conducts heat well | | | | | | | |
| conducts electricity well | | | | | | | |
| does not conduct heat well | | | | | | | |
| does not conduct electricity well | | | | | | | |
| is a metalloid | | | | | | | |
| is an alloy | | | | | | | |
| is an element | | | | | | | |
| looks shiny | | | | | | | |

**4. On the lines below, tell what a scanning tunneling microscope (STM) is and why it is useful.**

___________________________

___________________________

___________________________

___________________________

Name ____________________

Date ____________________

# Lesson 3—What Are the Properties of Some Common Substances?

## A. Classify

**Write the word *gas, liquid,* or *solid* on the line next to each situation.**

**1.** You skate on the ice in the winter. ____________

**2.** A woman who faints is given oxygen. ____________

**3.** You use a graphite pencil to spell your name. ____________

**4.** You drink milk with your lunch. ____________

**5.** A creek flows into the pond. ____________

**6.** Steam rises from the teapot. ____________

## B. Categorize

**In each group below, one word does not belong in the same category as the others. Circle the letter of that word. Then identify the remaining items as types of solids, liquids, or gases.**

**1.** a. oxygen
b. neon
c. hydrogen
d. carbon

____________________________________________

**2.** a. ice
b. wood
c. water
d. sand

____________________________________________

**3.** a. water
b. iron
c. mercury
d. blood

____________________________________________

Name ______________________

Date ______________________

**Study Skills**



# Lesson 3—What Are the Properties of Some Common Substances?

## Organize Information

**Graphic organizers can help you organize information and make sense of the facts you read.**

- Tables, charts, and webs are graphic organizers that can show main ideas and important details.
- Graphic organizers help you categorize, or group, information.
- Putting related ideas into categories makes it easier to find facts.

**As you read the lesson, continue to add new facts and details to the organizer.**

**States of Matter**

**Gases**

Air is a mixture of gases.

**Liquids**

Water is most common liquid on the planet.

**Solids**

Carbon is a common element on Earth.

**Molecules of Life**

Plant and animal life depend on carbon.

Name ______________________________

Date ______________________________

# Lesson 3—What Are the Properties of Some Common Substances?

1. **Investigation Skill Practice–Infer**

Bart tried to make an ice rink in his back yard by putting water into a wooden box. He made the box strong with long screws and thick wood. He was surprised that his box was broken in pieces when the water inside it froze. Make inferences to explain why the box broke.

______________________________________________________________

______________________________________________________________

______________________________________________________________

2. **Reading Focus Skill Practice–Main Idea and Details**

**Read the selection. Underline the main idea. Write 3 details on the lines below.**

Most of the objects and substances on Earth are made up of only a very few elements. Some elements, such as carbon, oxygen, and hydrogen are part of almost everything you touch. However some common elements are deadly to humans even in small amounts. For example, mercury was once used in thermometers, but it is very poisonous when inhaled or eaten. Lead is also a poison. It can cause children to become ill. Today it cannot be used in paints or other substances that people may be exposed to. Although oxygen is not a poison, ozone, made from three oxygen atoms is deadly to people who breathe in even a small amount.

______________________________________________________________

______________________________________________________________

______________________________________________________________

Name ____________________

Date ____________________

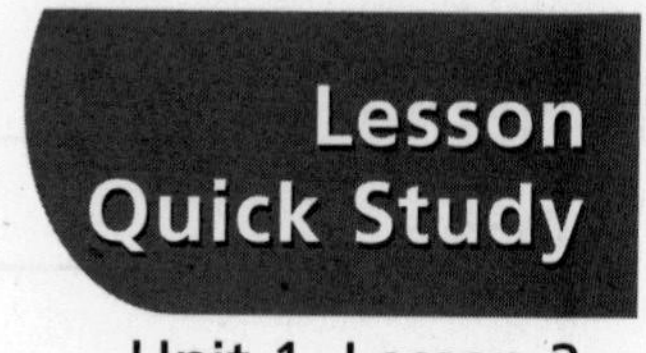




**Science Concepts**

**3. Recall concepts in the lesson by answering the questions about the states of matter.**

**Gases**

What are the two most important gases that make up our air?

____________________

Give an example of a noble gas.

____________________

**Liquids**

What happens to most liquids as they get cold?

____________________

What is the only metal that stays in a liquid state at room temperature?

____________________

**Solids**

Name several compounds that contain carbon.

____________________

What do charcoal, graphite, and diamonds have in common?

____________________

**Molecules of Life**

What are the five elements that make up DNA?

____________________

Name ____________________

Date ____________________

# Lesson 4—How Are Chemical and Physical Properties Used?

## A. Cloze Exercise

**Study the words and their definitions below. Then, using the underlined context clues, fill in the blanks with one of the words. Use all of the words once.**

**chemical property:** A property that involves the way a substance combines with other substances to form new substances.

**acid:** A chemical compound that turns blue litmus paper red and has a pH of less than 7

**base:** A chemical compound that turns red litmus paper blue and has a pH of more than 7

1. The pH of the substance is 8. This substance is a(n) ____________________.

2. An organism, such as pH of 4. My blueberry bushes grow well in a(n) ____________________ soil.

3. One of the ____________________ of the iron is its ability to combine with oxygen to make rust.

## B. Explore Word Meanings

**Answer the questions.**

1. Which of these pH readings are bases: 4.3, 8.5, 5.5, 7, and 7.5?

____________________

2. What would you call an ore: a chunk of pure gold or a chunk of rock with minerals that you cannot identify?

____________________

Name ________________________

Date ________________________

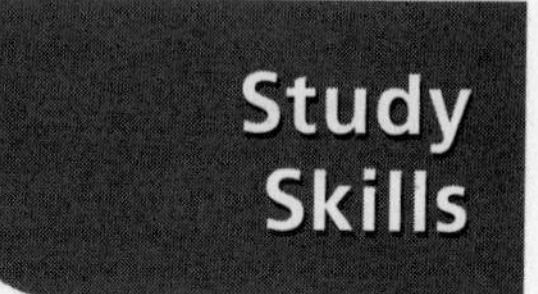




# Lesson 4—How Are Chemical and Physical Properties Used?

## Preview and Question

**Identifying main ideas in a lesson and asking questions about them can help you find important information.**

- To preview a lesson, read the lesson title and the section titles. Look at the pictures, and read their captions. Try to get an idea of the main topic and think of questions you have about the topic.
- Read to find the answers to your questions. Then recite, or say, the answers aloud. Finally, review what you have read.

**As you read this lesson, fill in the chart and practice reading, reciting, and reviewing.**

| How Are Chemical and Physical Properties Used? | | | | |
|---|---|---|---|---|
| **Preview** | **Questions** | **Read** | **Recite** | **Review** |
| Methods and reasons for separating mixtures | How does chemistry help scientists separate mixtures? | ✔ | ✔ | ✔ |
| | | | | |

Name ________________________

Date ________________________

# Lesson 4—How Are Chemical and Physical Properties Used?

1. **Investigation Skill Practice–Draw Conclusions**

**You hold a magnet over a mixture of metal shavings. Some of them are drawn to the magnet. Others stay on the table. What conclusions can you draw about the metal shavings mixture? What additional tests could you do to identify all the metals in the mixture?**

________________________________________

________________________________________

________________________________________

________________________________________

2. Focus Skill **Reading Focus Skill Practice–Main Idea and Details**

**Read the selection. Underline the main idea. Write two details on the lines below.**

The pH level of your garden soil is very important to know because some plants grow better at certain levels of pH. There are two ways to adjust the pH of your soil. First, if your soil is too acidic, you must add calcium carbonate, often called lime. The lime has a base pH and when you mix it with your soil, your soil will become less acidic. Second, if your soil is too basic, you should add organic matter. Organic matter could be compost, manure, leaves, sawdust, or peat moss.

________________________________________

________________________________________

________________________________________

Name ______________________________

## Science Concepts

3. **For each test below, explain how the test is conducted and what the test shows.**

**Solubility**

What does the test show?

______________________________

How do you conduct the test?

______________________________

**Litmus paper**

What does the test show?

______________________________

How do you conduct the test?

______________________________

**Flame test**

What does the test show?

______________________________

How do you conduct the test?

______________________________

**Melting point**

What does the test show?

______________________________

How do you conduct the test?

______________________________

Name ______________________________

Date ______________________________

# Lesson 4—How Are Chemical and Physical Properties Used?

## Acids, Bases, and Salts

### A. The pH Scale

**The pH scale shows how acidic or basic a solution is. It runs from 1, most acidic, to 14, most basic. Pure water is neither acidic nor basic and has a pH of 7.**

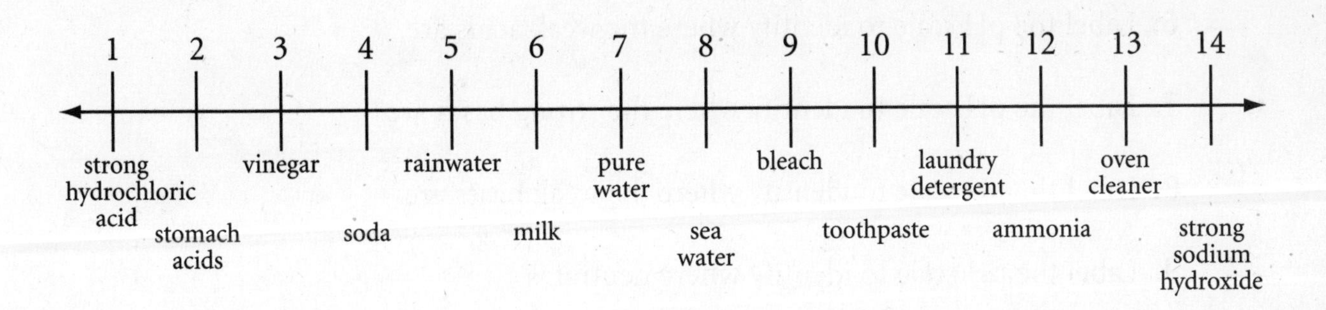

Name ______________________________

**Use the pH scale to help answer these questions.**

1. Is bleach an acid or base? ______________________________
2. Is soda an acid or base? ______________________________
3. Of rainwater, vinegar, and milk, which is the most acidic? __________
4. Of seawater, ammonia, and bleach, which is the most basic (alkaline)? ______
5. Label the pH scale to identify where the strong acids are.
6. Label the pH scale to identify where the weak acids are.
7. Label the pH scale to identify where the strong bases are.
8. Label the pH scale to identify where the weak bases are.
9. Label the pH scale to identify where neutral is.

Name ________________________________

## B. Salts

**The "salt" you add to food is just one example of a group of substances called salts. Salts form when a strong acid reacts with a strong base. Water can also form.**

In this example, nitric acid and potassium hydroxide react to form potassium nitrate salt and water.

$$HNO_3 + KOH \rightarrow KNO_3 + H_2O$$

**Look at how the elements on the left side of the reaction recombine to form the substances on the right.**

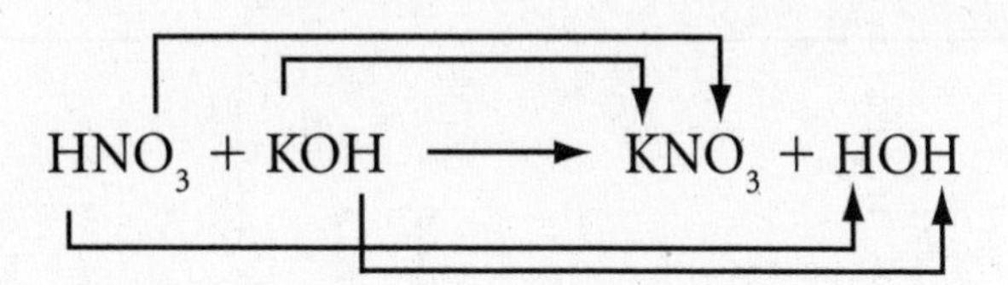


The next example looks more complicated, but it still shows a strong acid and strong base making a salt and water. Sulfuric acid and potassium hydroxide react to form potassium sulfate salt and water. The same elements on the left end up on the right, but in different combinations.

$$H_2SO_4 + KOH \rightarrow K_2SO_4 + H_2O$$

**1.** What do you need to make a salt? ________________________________

**2.** What happens when a strong acid and a strong base react? ________________

________________________________________________________________

**3.** If you combine stomach acid and oven cleaner, would the outcome be a salt and water? Explain. ________________________________________________

________________________________________________________________

________________________________________________________________

Name ______________________

4. If you combine water and bleach, would the outcome be a salt and water? Explain. ______________________

______________________

5. Look at the pH scale. Identify two products that will likely combine to form a salt and water. ______________________

6. Predict the formula for the salt that forms when the strong acid hydrochloric acid (HCl) and the strong base potassium hydroxide (KOH) react. Try to write the equation that represents the reaction.

______________________

Name ______________________________

Date ______________________________

Unit 1, Lesson 5

# Lesson 5—What Are Chemical Reactions?

## A. Word Families

**Fill in the blanks with the correct form of the word.**

**1.** reactant/react
Noun: One ____________ in the experiment was sodium. It combined with chlorine to make table salt.
Verb: The elements iron and oxygen ____________ with each other in the presence of water to form rust.

**2.** salt/salty
Noun: When a base and an acid come in contact, the result is some type of ____________.
Adjective: The ocean water left a ____________ taste after it dried from my skin.

**3.** product/produce
Noun: What ____________ do you get from the combination of sodium and chlorine?
Verb: The chemicals ammonia and bleach combine to ____________ a terrible, poisonous gas.

**4.** chemical reaction/chemically reacting
Adjective and noun: Sunlight, water, and air create a ______________________ in a plant to make its food.
Adverb and verb: You can tell that the substances are ______________________ because they create bubbles and turn a different color.

## B. Write Sentences

**Write your own sentences using at least two words from the box.**

| chemical reaction | reactant | product | salt |
|---|---|---|---|

______________________________________________________________

______________________________________________________________

______________________________________________________________

______________________________________________________________

Name ______________________________

Date ______________________________

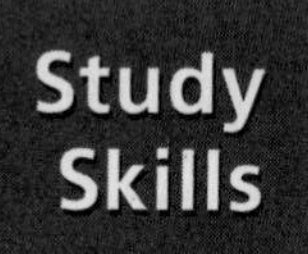




# Lesson 5—What Are Chemical Reactions?

## Skim and Scan

**Skimming and scanning are two ways to learn from what you read.**

- To skim, quickly read the lesson title and the section titles. Look at the visuals, or images, and read the captions. Use this information to identify the main topics.
- To scan, look quickly through the text for specific details, such as key words or facts.

**Before you read this lesson, skim the text to find the main ideas. Then look for key words. If you have questions about a topic, scan the text to find the answers. Fill in the chart below as you skim and scan.**

| What Are Chemical Reactions? | |
|---|---|
| **Skim** | **Scan** |
| Lesson Title: | Key Words and Facts: |
| Main Idea: | |
| Section Titles: | |
| Visuals: | |

Name ______________________________

Date ______________________________

# Lesson 5—What Are Chemical Reactions?

1. **Investigation Skill Practice–Develop a Testable Question**

**Juan noticed that when he made pancakes with baking powder, they were fluffy and fat. When he forgot to add baking powder to the pancake batter, his pancakes were flat and dense. What testable question could Juan develop about this observation? What procedure could he use to test his question?**

______________________________

______________________________

______________________________

______________________________

______________________________

2. Focus Skill **Reading Skill Practice–Compare and Contrast**

**Read the selection. Compare and contrast the ice on the steps and on the sidewalk.**

Roadway ice is dangerous for motorists. Not all towns use the same methods for melting it. In some places, towns put salt on the roads. The salt makes the freezing temperature of water change. That way the roads stay wet instead of icy. However, salt can harm plants when it runs off the road. In other places, towns put down sand instead of salt. Sand is usually dark brown or black. The sand works with the sun to absorb heat and melt the ice. It also helps tires grip the slippery ice. The sand does not harm plants that grow near the streets.

______________________________

______________________________

______________________________

______________________________

______________________________

Name ______________________

## Science Concepts

**3. Identify each of the following as a physical change or a chemical change. Write *physical change* or *chemical change* on the line.**

water turns to ice ______________________
a tomato rots in the refrigerator ______________________
sugar dissolves in water ______________________
an iron gate becomes rusty ______________________

**4. Identify the reactants and the products in the following reactions.**

Sodium + Chlorine = sodium chloride (table salt)
What are the reactants? ______________________
What is the product? ______________________

iron + oxygen = rust (iron oxide)
What are the reactants? ______________________
What is the product? ______________________

**5. When nitrogen dioxide is heated it splits into two elements.**

What elements does it become? ______________________
How does the mass of the nitrogen dioxide compare to the mass of the products of the chemical reaction?

______________________

**6. When you combine the elements hydrogen and oxygen you get $H_2O$.**

What compound have you created? ______________________
How does the mass of the hydrogen and the oxygen compare to the mass of the $H_2O$ that is the product of the chemical reaction?

______________________

**7. Answer the questions about salt.**

What are the physical properties of salts?

______________________

How are salts formed?

______________________

Name ___________________________

Date ___________________________

# Lesson 5—What Are Chemical Reactions?

## Chemical Reactions

Physical properties can be observed without changing the type of substance. They include density and melting point. Chemical properties describe how a substance reacts with other substances, such as how it burns.

Physical changes do not produce new substances. They can be easily undone. Examples include folding paper or melting ice cubes. Chemical changes do produce new substances. They are difficult to undo. Examples include burning paper or adding acid to metal.

**A. Consider each property in this chart. Check one column to show if it is a physical or chemical property.**

| Property | Physical? | Chemical? |
|---|---|---|
| 1. texture | | |
| 2. ability to rust | | |
| 3. volume | | |
| 4. how it reacts to acid | | |
| 5. boiling point | | |
| 6. color | | |

**B. Consider each change in this chart. Check one column to indicate if it is a physical or chemical change.**

| Change | Physical? | Chemical? |
|---|---|---|
| 1. carving wood into a baseball bat | | |
| 2. dissolving sugar in water | | |
| 3. burning a candle | | |
| 4. freezing water into ice cubes | | |
| 5. metal nail rusting | | |
| 6. tearing paper | | |

Name ______________________________

### C. Iodine-Starch Reactions

The pictures below show the steps and results of an activity about iodine and starch. You will use it to answer the questions. You can also perform the investigation, if you have the proper materials and supervision.

**ACTIVITY:** Potatoes and Iodine

You will need:

- potato
- knife
- iodine solution
- eyedropper

If you place a few drops of iodine solution on the cut face of a potato, the solution will turn dark purple-black.

**Identify each change in the investigation as physical or chemical:**

**1.** cutting the potato ______________________________

**2.** iodine solution changing color ______________________________

**3.** Potatoes are starchy foods. What does the color change in iodine indicate? ______________________________

Name ______________________________
Date ______________________________

# Lesson 1—How Do Organisms Transport Materials?

## A. Words with Multiple Meanings

**Think about how the meaning of the underlined word is used in each sentence. Circle the correct definition.**

**1.** Marie used a microscope to compare a plant cell and an animal cell.

small, isolated room

basic unit of life

**2.** Nervous tissue carries signals from the brain to parts of the body.

cells that carry out a certain function

sheet of absorbent paper

**3.** The heart is a hard-working organ that pumps blood to all parts of the body.

musical instrument with sets of pipes controlled by a keyboard

group of tissues working together.

## B. Content Area Words

**Write the letter of the definition that matches the word.**

**1.** ______ capillaries

**2.** ______ organ system

**3.** ______ nucleus

**4.** ______ cytoplasm

**5.** ______ organelles

**6.** ______ vacuoles

**a.** a group of organs that work together to do a job for the body

**b.** structures within a cell with a specific function for keeping the cell alive

**c.** the organelle that directs all of a cell's activities

**d.** the jelly between the cell membrane and the nucleus

**e.** organelles that store nutrients, water, or waste materials until the cell uses them or gets rid of them

**f.** tiny blood vessels

Name ______________________

Date ______________________

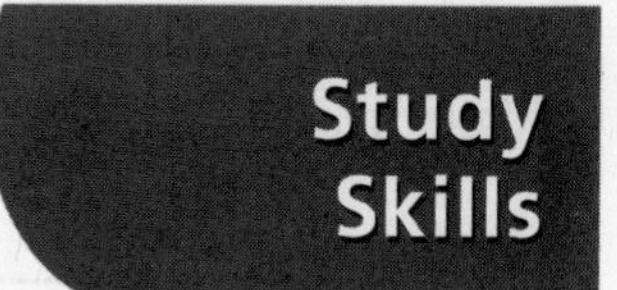




# Lesson 1—How Do Organisms Transport Materials?

## Anticipation Guide

**An anticipation guide can help you anticipate, or predict, what you will learn as you read.**

- Look at the section titles for clues.
- Preview the Reading Check question at the end of each section. Use what you know about the subject of each section to predict the answers.
- Read to find out whether your predictions were correct.

**As you read each section, complete the anticipation guide below. Predict answers to each question and check to see if your predictions were correct.**

| How Do Organisms Transport Materials? | | |
|---|---|---|
| **Cells as Building Blocks** | | |
| **Reading Check** | **Prediction** | **Correct?** |
| How do cells keep organisms alive and healthy? | In this section, we will learn about the function of cells. | |
| **Cell Structures and Functions** | | |
| **Reading Check** | **Prediction** | **Correct?** |
| What organelle directs all the functions of a cell? | | |
| **Cells, Systems, Organs, Systems** | | |
| **Reading Check** | **Prediction** | **Correct?** |
| What life functions do tissues and organs carry out? | | |
| **Transport in Multicellular Organisms** | | |
| **Reading Check** | **Prediction** | **Correct?** |
| What are the functions of transport tissues in plants and animals? | | |

Name ______________________________

Date ______________________________

# Lesson 1—How Do Organisms Transport Materials?

**1. Investigation Skill Practice–Classify**

**Look at the list of tissues in the box. Classify them in the chart under the correct type of tissue.**

| blood | cartilage | skin | the lining of your capillaries |
|---|---|---|---|
| bones | muscles | spinal cord | |
| brain | nerves | | tongue muscle |

| muscle tissue | epithelial tissue | connective tissue | nervous tissue |
|---|---|---|---|
| | | | |

**2. Focus Skill — Reading Focus Skill Practice–Main Idea and Details**

**Read the selection. Underline the main idea. Write 3 details on the lines below.**

Each system of the body is dependent on the others, but the circulatory system is the one that gets around the most. For example, the circulatory system gathers waste from each cell and delivers it to the excretory system. That's how wastes are eliminated from the body. It works with the respiratory system by collecting and distributing oxygen to every part of the body. It works with the digestive system by gathering nutrients from food and supplying them to all the organs and tissues.

______________________________________________________________

______________________________________________________________

______________________________________________________________

______________________________________________________________

Name ______________________

## Science Concepts

**3. Read the list of details explaining how an organism transports materials. Write each detail that applies to plants under *Plant Cells.* Write each detail that applies to animals under *Animal Cells.* Some may apply to both.**

**Details**

Cells contain a nucleus, cell membrane, and cytoplasm.

The transport system includes the heart, blood vessels, and blood.

The transport system carries water and minerals from roots to leaves.

The transport system helps carryout the removal of wastes.

The transport system helps deliver oxygen throughout the organism.

Organisms have specialized tissues that serve different functions.

Cells work together to carry out life functions.

| Animal Cells | Plant Cells |
|---|---|
| | |
| | |
| | |
| | |
| | |
| | |
| | |
| | |
| | |

**4. Name two more functions of plant and animal cells that were not on the list of details above.**

______________________________________________

Name ______________________________

Date ______________________________

# Lesson 2—How Do the Circulatory and Respiratory Systems Work Together?

## A. Word Origins

**Fill in the blank with the word from the box that fits the word origin. Use a dictionary if you need help.**

| | | |
|---|---|---|
| circulatory system | arteries | veins |
| respiratory system | capillaries | |

| Word Origin | Language | Meaning | Vocabulary Word |
|---|---|---|---|
| 1. *arteria* | Greek | to lift, take up | ______________ |
| 2. *capillus* | Latin | hair | ______________ |
| 3. *vena* | Latin | vein | ______________ |
| 4. *circulatus* | Latin | to make a circle | ______________ |
| 5. *respirare* | Latin | of breathing | ______________ |

## B. Context Clues

**Use context clues to complete each sentence correctly.**

1. The blood vessels that carry blood away from the heart are

   ______________ .

2. Breathing smog, pollution, or smoke can cause problems in your

   ______________ .

3. Blocked or hardened blood vessels put a burden on a person's

   ______________ .

4. The heart pumps the blood that returns from the body's organs

   through large ______________ .

Name ______________________

Date ______________________

**Study Skills**



# Lesson 2—How Do the Circulatory and Respiratory Systems Work Together?

## Pose Questions

**Posing, or asking, questions as you read can help you understand what you are reading.**

- Form questions as you read. For example, you may ask how a science concept is connected to other concepts.
- Use the questions to guide you reading. Look for answers as you read.

**Before you read this lesson, write a list of questions in the chart below. Look for the answers as you read. Record the answers in the chart.**

| How Do the Circulatory and Respiratory Systems Work Together? | |
|---|---|
| **Questions** | **Answers** |
| How does the air I breathe get into my blood? | Tiny air sacs in the lungs pass the oxygen to the blood through the capillaries, or tiny blood vessels. |
| ______________________<br>______________________ | ______________________<br>______________________ |
| ______________________<br>______________________ | ______________________<br>______________________ |
| ______________________<br>______________________ | ______________________<br>______________________ |

Name ______________________________
Date ______________________________

# Lesson 2—How Do the Circulatory and Respiratory Systems Work Together?

## 1. Investigation Skill Practice–Identify the Dependent and Controlled Variables

Janet wants to determine how different people respond to exercise. She chooses one athlete from the track team, one athlete from the golf team, and one person who does not regularly exercise. She tracks their heart rate three times: once sitting, once after walking for five minutes, and once after running for five minutes.

**What is the controlled variable in Janet's experiment?**

______________________________________________________________

**What is the dependent variable in Janet's experiment?**

______________________________________________________________

## 2. Focus Skill Reading Focus Skill Practice–Sequence

**Put the following events about the respiratory system in the correct sequence. Number the steps 1 to 4.**

______ Clean air travels down your trachea, into smaller and smaller tubes into your lungs.

______ Carbon dioxide leaves your body when you exhale.

______ Carbon dioxide passes from the blood plasma into the air sacs in the lungs while oxygen passes from the air sacks into the blood.

______ You breathe air in through your nose.

Name ______________________________

## Science Concepts

**3. Read the events that occur in the relationship between the circulatory and the respiratory systems below. Write them in the correct order in the flow chart.**

The heart pumps the returned blood to the lungs.

Oxygen moves into capillaries from the lungs.

Air is breathed in through the nose or mouth.

Capillaries allow oxygen to move into cells and waste materials to move out.

Oxygen-rich blood goes to the heart.

The heart pumps blood out through the arteries.

Blood flows from the arteries into capillaries.

Blood is returned to the heart through veins.

Air travels through the trachea into the lungs.

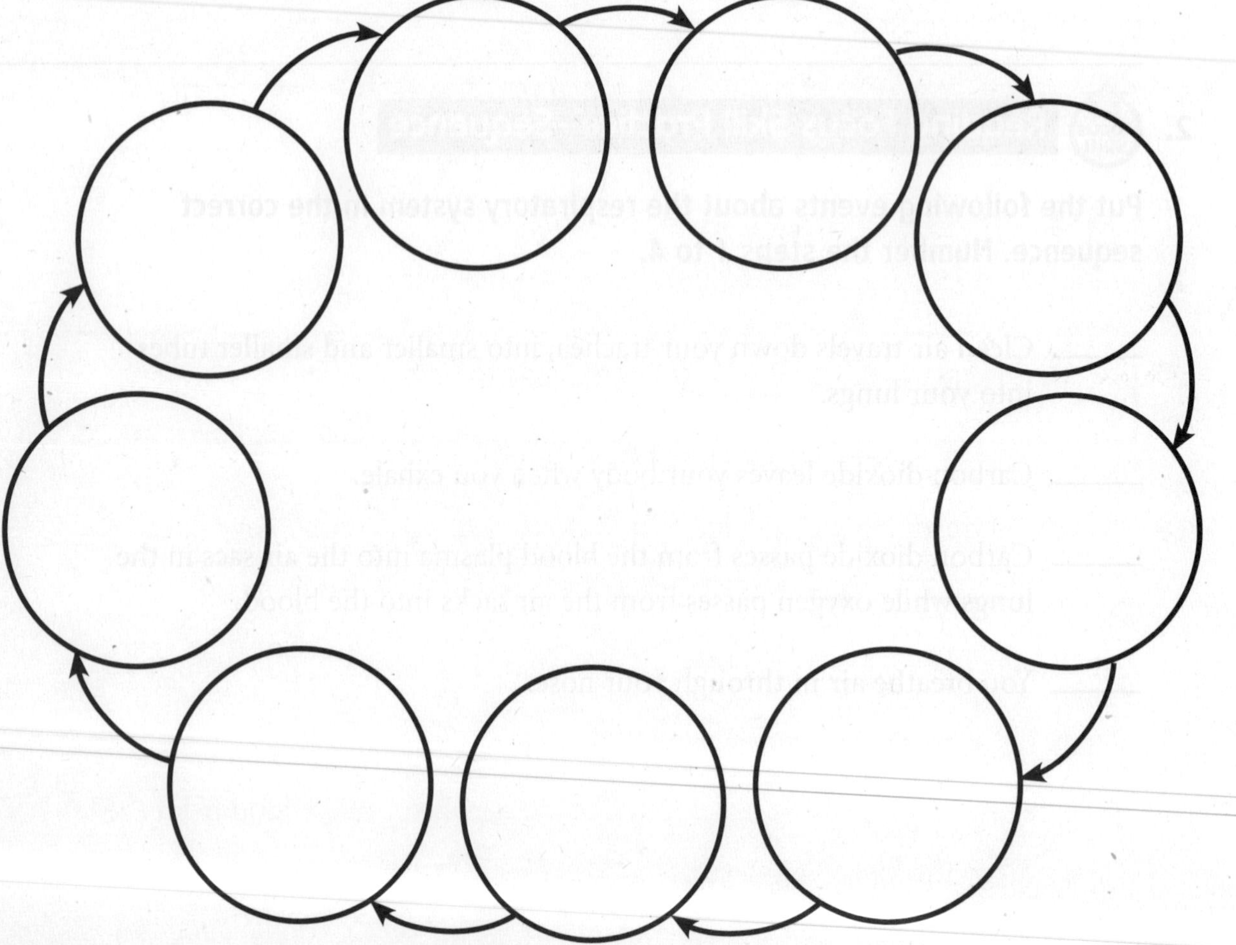

Name ______________________________

Date ______________________________

# Lesson 3—How Do the Organs of the Digestive System Work Together?

## A. Word Origins

**Learning the origins of a word sometimes helps you remember its meaning. Fill in the table with the correct vocabulary word from the box.**

| digest | esophagus | peristalsis | villi | intestine |
|---|---|---|---|---|

| Word Origin | Language | Meaning | Vocabulary Word |
|---|---|---|---|
| *oisophagos* | Latin | passage for food | 1. |
| *peristaltikos* | Greek | to surround | 2. |
| *vellus* | Latin | shaggy hair | 3. |
| *digestus* | Latin | to arrange | 4. |
| *intestinus* | Latin | inward, internal | 5. |

## B. Context Clues

**Complete the sentence with the correct word or phrase from the box.**

| digestive system | esophagus | peristalsis |
|---|---|---|

**1.** If you are an athlete, you know how important it is to keep your ________________ healthy by eating lots of fruits and vegetables.

**2.** Toys with small parts are a choking hazard because they can get stuck in a child's ________________.

**3.** The movement called ________________ helps move food from the stomach through the intestines to the colon.

Name ______________________________

Date ______________________________

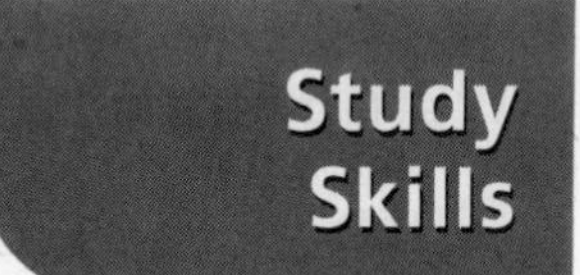




# Lesson 3—How Do the Organs of the Digestive System Work Together?

## Use Visuals

**Visuals can help you better understand and remember what you read.**

- Photographs, illustrations, diagrams, charts, and maps are different kinds of visuals.
- Many visuals have titles, captions, or labels that help readers understand what is shown.
- Visuals often show information that appears in the text, but they show it in a different way.

**As you read this lesson, look closely at the visuals and the text that goes with them. Answer the questions in the checklist.**

| How Do the Organs of the Digestive System Work Together? | |
|---|---|
| | What kind of visual is shown to illustrate the digestive system?<br>______________________________<br>______________________________ |
| | What does the visual show?<br>______________________________<br>______________________________ |
| | How does the visual relate to the lesson you are reading?<br>______________________________<br>______________________________ |
| | How does the visual help you better understand the subject of what you are reading?<br>______________________________<br>______________________________ |

Name ______________________

Date ______________________

# Lesson 3—How Do the Organs of the Digestive System Work Together?

1. **Investigation Skill Practice–Observe, Record Data, and Make Inferences**

**Jill covered the inside of a strainer with a thin cloth. Then she placed a scoop of plain yogurt over the cloth. She watched and took note of what happened every ten minutes for one hour. She noticed that the liquid from the yogurt dripped out through the cloth into a bowl below.**

**What did Jill observe?**

______________________________________________

**When did Jill record her data?**

______________________________________________

**What inferences do you think Jill could make about how nutrients move from the intestines into the blood?**

______________________________________________

2. Focus Skill **Reading Skill Practice–Sequence**

**Put the following events about the process of digestion in the correct sequence. Number the steps 1 to 5.**

_____ Food is churned and mixed with chemicals that break it down.

_____ Food travels to the stomach through the esophagus.

_____ Chewing breaks food down.

_____ Broken down food travels out of the stomach to the intestine.

_____ Saliva in the mouth begins digestion of starches.

Name ______________________________

## Science Concepts

**3. Events in a Sequence**

**Tell what happens in each organ of the digestive system listed below.**

**mouth**

______________________________

**esophagus**

______________________________

**stomach**

______________________________

**small intestine**

______________________________

**large intestine**

______________________________

**colon**

______________________________

**4. How does the digestive process start in your mouth?**

______________________________

______________________________

Name ______________________________

Date ______________________________

# Lesson 4—How Do Plants and Animals Rid Themselves of Wastes?

## A. Graphic Organizer

**Fill in the blanks of the graphic organizer with details from the definition of each vocabulary word.**

| Vocabulary Word | Plants or Animals? | Function in Eliminating Wastes |
|---|---|---|
| urea | | |
| nephrons | | |
| stomata | | |
| ureters | | |
| transpiration | | |
| kidney | | |

## B. Context Clues

**Write the word or phrase from the box that best completes each sentence.**

| transpiration | kidneys | excretory system |
|---|---|---|

1. When the woman's ________________ failed, she began treatment to have her blood filtered by a machine.
2. The plant's ability to perform ________________ means that it can eliminate water through holes in its leaves.
3. Your body stays healthy because its wastes are removed by your ________________.

Name ______________________________

Date ______________________________

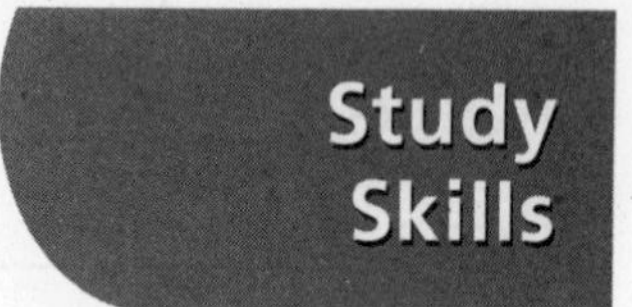




# Lesson 4—How Do Plants and Animals Rid Themselves of Wastes?

## Make an Outline

**An outline is a good way to record main ideas and details.**

- Topics in an outline are shown by Roman numerals.
- Main ideas about each topic are shown by capital letters.
- Details about each main idea are identified by numbers.

**As you read this lesson, remember to pay attention to the topics, main ideas, and details. Use the information to complete the outline below.**

How Do Plants and Animals Rid Themselves of Wastes?

I. The Excretory System

A. Wastes are removed by the excretory system.

1. ______________________________

2. ______________________________

B. The excretory system is made up of kidneys, ureters, bladder, and urethra.

1. ______________________________

2. ______________________________

3. ______________________________

4. ______________________________

II. How Plants Rid Themselves of Waste

A. ______________________________

1. ______________________________

2. ______________________________

B. ______________________________

1. ______________________________

2. ______________________________

Name ______________________________

Date ______________________________

# Lesson 4—How Do Plants and Animals Rid Themselves of Wastes?

## 1. Investigation Skill Practice–Record Data, Make Inferences

Louis wondered how different plants eliminate waste. He studied the life cycles of many plants, such as tulips, pine trees, maple trees, palm trees, and house plants. He noticed that some plants drop leaves during cool weather. Others die back after blooming. Some do not drop leaves at all and never appear to die.

What data should Louis record as he makes his observations?

______________________________________________________________

______________________________________________________________

What inferences can Louis make about how different plants rid themselves of waste?

______________________________________________________________

______________________________________________________________

## 2. Focus Skill Reading Focus Skill Practice–Compare and Contrast

**Read the selection. Compare and contrast a dialysis machine and the kidneys.**

When a person's kidneys do not function properly, he or she may go on dialysis. The dialysis machine works like a kidney. It forces all of the body's blood through a filter. It also separates waste material as a kidney would do. Although its function is similar, the experience is much different. While kidneys work all day and night, a dialysis machine filters the blood only three times each week. With kidneys, blood stays inside the body to be filtered. On a dialysis machine, blood must be taken out of the body, filtered, and then put back in. The procedure is uncomfortable, but for many, it is their only hope for life.

______________________________________________________________

______________________________________________________________

______________________________________________________________

______________________________________________________________

Name ______________________

## Science Concepts

**3.** Answer the questions in the chart for animals and plants. Be as specific as possible. Use your book if you need help.

| | **Animals** | **Plants** |
|---|---|---|
| What wastes are produced? | | |
| Where are wastes stored? | | |
| How are wastes eliminated? | | |
| What are the parts of the excretory system? | | |

**4.** Answer the questions about how animals and plants eliminate wastes.

What is the function of the kidneys in animals?

______________________

______________________

How do the stomata help a plant eliminate waste?

______________________

______________________

Name ______________________

Date ______________________

# Lesson 5—How Are Materials Transported in Vascular Plants?

## A. Word Origins/Etymology

**Write the word from the box that fits the word origin.**

| xylem | phloem | vascular tissue |
|---|---|---|

1. A word that derives from *xylon*, the Greek word for "wood."

   ______________

2. A word that derives from *phloos*, the Greek word for "bark."

   ______________

3. A word that comes from *vasculum*, the Latin word for "small vessel."

   ______________.

## B. Explore Word Meanings

**For each description below, write *vascular plant* or *nonvascular plant*.**

1. A plant without roots ______________
2. A plant with a taproot ______________
3. A plant without leaves or stems ______________
4. A plant that can carry nutrients to each part ______________
5. A plant that grows a ring of xylem each year ______________

Name ______________________________

Date ______________________________

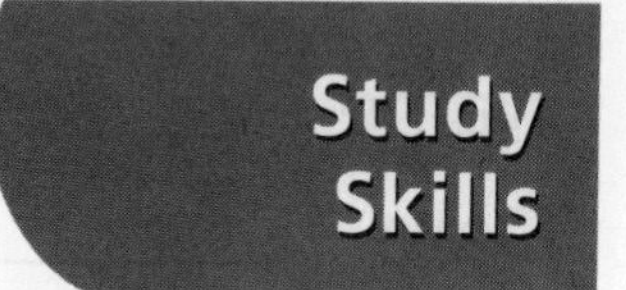




# Lesson 5—How Are Materials Transported in Vascular Plants?

## Connect Ideas

**You can use a web organizer to show how different ideas and information are related.**

- List important themes in the ovals in the web's center.
- Add ovals showing main ideas that support each theme.
- Add bubbles for the details that support each main idea.

**Complete the web below as you read this lesson. Fill in each bubble by adding facts and details. Add more bubbles if you need them.**

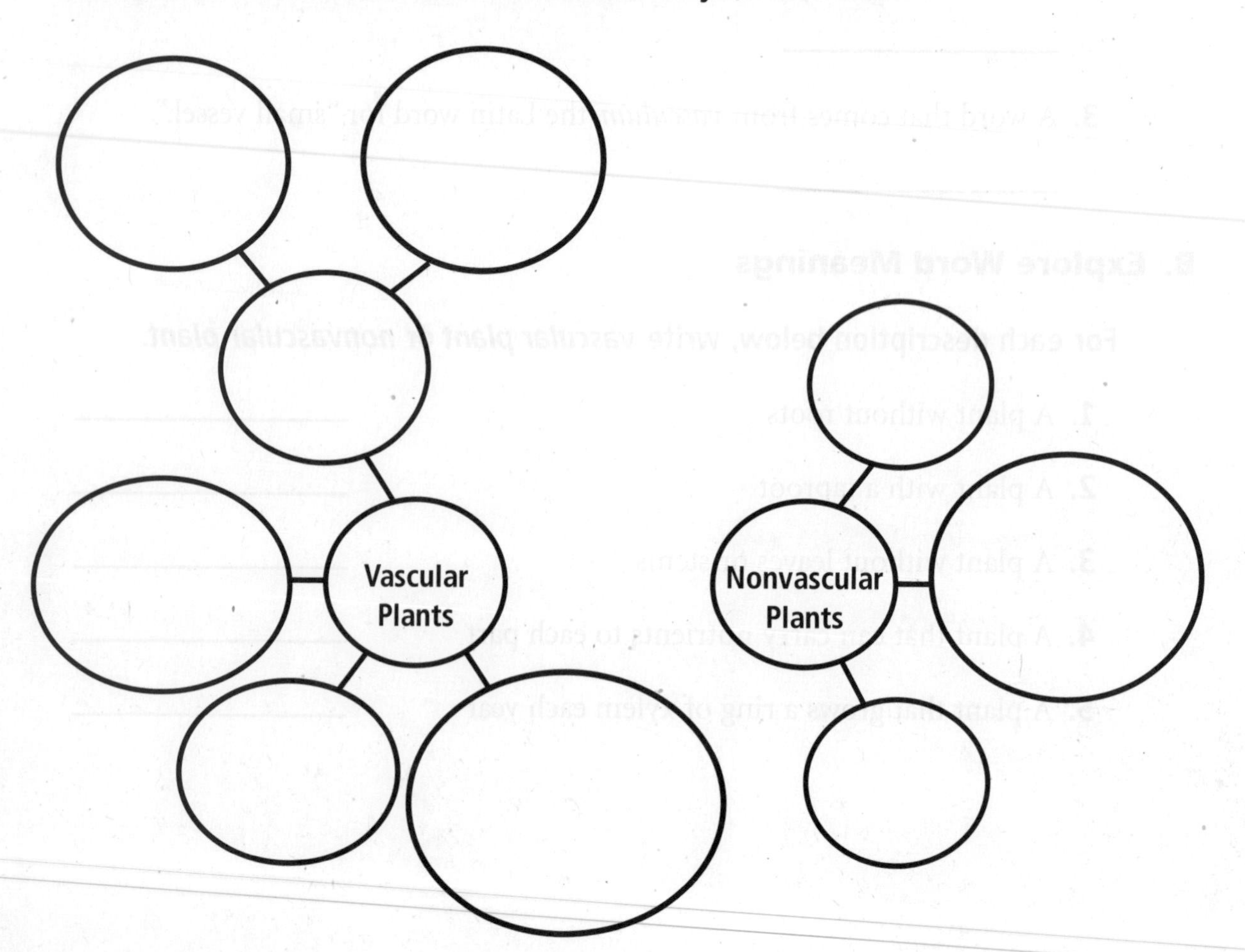

Name ______________________

Date ______________________

# Lesson 5—How Are Materials Transported in Vascular Plants?

1. **Investigation Skill Practice–Infer**

In early spring, when the sap begins to flow, a farmer puts her taps in the trunks of her sugar maple trees. These taps allow a thin sweet sap to drip slowly into a bucket. The farmer empties the buckets every day. After several weeks, she has collected a great cauldron of sap. She boils down the sap to make real maple syrup. What can you infer about the maple syrup process?

What part of the tree do the taps go into?

______________________

______________________

Why is the sap sweet?

______________________

______________________

2. **Reading Focus Skill Practice–Main Idea and Details**

**Read the selection. Underline the main idea. Write 3 details on the lines below.**

Like any plant, a grape vine has specific needs that must be met to make it thrive. First, grape vines require very deep soil because they have long roots that reach far below the surface of the earth. Second, grape vines grow best in places with a long, warm growing season. This allows the plant to set fruit and ripen before it gets too cold. Finally, grapes need good air circulation, such as a windy spot on a hill. Without good circulation, leaves cannot release their moisture, and the plant and fruit may rot before it gets ripe.

______________________

______________________

______________________

______________________

Name ______________________

## Science Concepts

**3.** Read the questions below. Write the word *leaves, stems,* or *roots* to answer the question. Some questions will have more than one answer.

What parts do vascular plants have that nonvascular plants do not?

______________________

What part uses sunlight, air and water to make sugar?

______________________

What part of the plant contains phloem that carries the sugar to the rest of the plant?

______________________

What part provides support for a plant?

______________________

Where does the vascular tissue gather in rings that show the age of a tree?

______________________

What part absorbs water and nutrients from the soil?

______________________

What parts of the plant contains vascular tissue?

______________________

What part has chloroplasts for making food?

______________________

What part anchors the plant to the ground?

______________________

**4.** What is the function of xylem and phloem? Write your explanation on the lines below.

______________________

______________________

______________________

Name ______________________________

Date ______________________________

# Lesson 6—How Do Cells Get the Energy They Need?

## Cloze Exercise

**Study the words and their definitions below. Then, using context clues, fill in the blanks with one of the words. Use all of the words once.**

**photosynthesis:** the process by which plants make food from carbon dioxide and water and release oxygen into the air

**cellular respiration:** the process by which cells use oxygen to break down sugar to release energy

**fermentation:** the process that releases energy from sugar in the absence of oxygen

**chlorophyll:** green pigment in a plant that allows a plant cell to use light to make food

In the fall, when the trees loose their green pigment, the leaf cells lose their

______________________.

The process of ______________________ occurs during yogurt production in the absence of oxygen.

You might say that the energy you feel comes from the break down of sugar called ______________________.

Animals depend on ______________________ because the process releases oxygen to the air for animals to breathe.

Name ______________________________

Date ______________________________

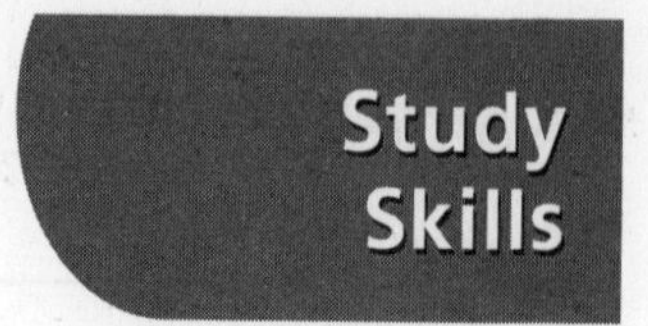




# Lesson 6—How Do Cells Get the Energy They Need?

## Take Notes

**Taking notes can help you remember important ideas.**

- Write down important facts and ideas. Use your own words. You do not have to write in complete sentences.
- One way to organize notes is in a chart. Write down the main ideas in one column and facts and details in another.

**As you read this lesson, use the chart below to take notes.**

| How Do Cells Get the Energy They Need? | |
|---|---|
| **Main Ideas** | **Facts** |
| • Photosynthesis is the process that makes food for the plant. | • The process of photosynthesis supplies enough oxygen into the air for humans to breathe. |
| • ______________________ | • ______________________ |
| • ______________________ | • ______________________ |
| • ______________________ | • ______________________ |
| • ______________________ | • ______________________ |

Name ______________________________

Date ______________________________

# Lesson 6—How Do Cells Get the Energy They Need?

1. **Investigation Skill Practice–Identify the Independent Variable**

Kaylee conducted an experiment with two identical plants. She planted both plants in rich soil. She put both plants in a sunny window. To one plant, she gave three tablespoons of water three times each day. To the other plant, she gave one tablespoon of water once each day. She observed how both plants grew.

What is Kaylee studying in her experiment?

______________________________________________________________

______________________________________________________________

What is the independent variable in Kaylee's experiment?

______________________________________________________________

______________________________________________________________

2. Focus Skill **Reading Skill Practice–Sequence**

**Put the following events about the process of photosynthesis and cellular respiration in the correct sequence. Number the steps 1 to 5.**

_____ Plants use carbon dioxide and water to make sugar during photosynthesis.

_____ An animal breathes oxygen from the air.

_____ Water and carbon dioxide are released as waste.

_____ Oxygen and water are released as waste.

_____ Oxygen helps turn sugars to energy in the mitochondria.

Name ___________________________

## Science Concepts

### 3. Identify the Sequence

**The events of photosynthesis are similar to the events in cellular respiration. Read the events listed in the chart and copy each event under *Photosynthesis* or *Cellular Respiration.* Put the events in the correct order under each heading.**

| |
|---|
| Carbon dioxide and water are released into the blood. |
| Carbon dioxide combines with hydrogen from the water to make glucose. |
| Carbon dioxide is exhaled through the lungs. |
| Glucose is used for food. |
| Oxygen in the mitochondria breaks down sugar. |
| Oxygen and water is released into the air. |
| Sunlight is absorbed by chloroplasts. |
| The energy from the sun causes water molecules to split. |
| Water is carried to the kidneys. |
| When sugar breaks down, it releases energy. |

**Photosythesis**

**Cellular Respiration in Animal Cells**

Name ______________________

Date ______________________

**Extra Support**

Unit 2, Lesson 6

# Lesson 6—How Do Cells Get the Energy They Need?

## A. Photosynthesis and Respiration

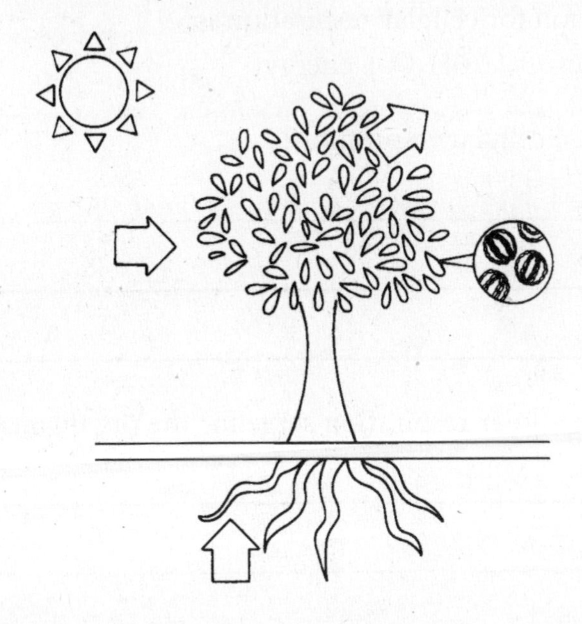

**The picture above shows how a plant captures energy and stores it in the process of photosynthesis. Describe how each word below is a part of photosynthesis. You may use drawings and labels as a part of your description.**

**1.** oxygen

______________________________

**2.** water

______________________________

**3.** sunlight

______________________________

**4.** carbon dioxide

______________________________

Name ______________________

## B. Respiration Equation

**Use the equation and what you already know about cellular respiration to answer the questions below.**

The chemical equation for cellular respiration is:

$C_6H_{12}O_6 + 6O_2 \rightarrow 6CO_2 + 6H_2O + \text{energy}$

**1.** Which organisms use cellular respiration?

______________________

______________________

______________________

**2.** What purpose does cellular respiration serve for the organisms that perform it?

______________________

______________________

______________________

**3.** Use the chemical equation above to write a series of steps that describes what is happening in the process of cellular respiration. Be sure to tell which materials are taken in and which materials are released as part of the process.

______________________

______________________

______________________

______________________

______________________

**4.** What happens to the waste products of respiration in the human body?

______________________

______________________

______________________

Name ______________________________

Date ______________________________



# Lesson 1—How Does Water Move from Earth to the Air and Back Again?

## A. Explore Word Meanings

**Match the word on the left to its definition on the right.**

**1.** ___ evaporation

**2.** ___ hydrologic cycle

**3.** ___ condensation

**4.** ___ fog

**5.** ___ water cycle

**6.** ___ glaciers

**7.** ___ water vapor

**a.** a word for the water cycle

**b.** huge sheets of ice

**c.** a cloud that forms near the ground

**d.** the constant movement of water from Earth's surface to the atmosphere and back to Earth's surface

**e.** the gas form of water

**f.** the process by which a liquid changes into a gas

**g.** the process by which a gas changes into a liquid

## B. Context Clues

**Complete the sentence with a word from the box.**

| water vapor | condensation | evaporation |
|---|---|---|

**1.** The night temperature was freezing, so ______________ in the air inside the car froze to the inside of the car windows.

**2.** We turned on the heat, and as the windows warmed up, the ice melted and gathered as wet drops of ______________ on the windows.

**3.** Soon the car was toasty warm inside which allowed ______________ to take place. The windows were dry again.

Name ____________________

Date ____________________

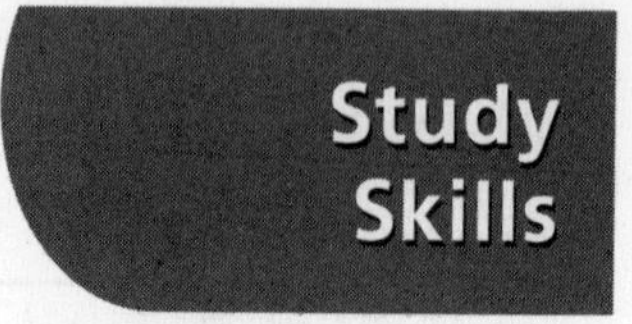




# Lesson 1—How Does Water Move from Earth to the Air and Back Again?

## Use a K-W-L Chart

**A K-W-L chart can help you focus on what you already know about a topic and what you want to learn about it.**

- Use the K column to list what you already know about the water cycle.
- Use the W column to list what you want to learn about the water cycle.
- Use the L column to list what you have learned about the water cycle from your reading.

**Complete your own K-W-L chart as you read this lesson.**

| How Does Water Move from Earth to the Air and Back Again? | | |
|---|---|---|
| **What I Know** | **What I Want to Learn** | **What I Learned** |
| • I know that ocean water is salty and cannot be used for drinking. | • How does it get salty?<br>• Where does fresh water come from? | • ____________ |
| • ____________<br>• ____________<br>• ____________ | • ____________<br>• ____________<br>• ____________ | • ____________<br>• ____________<br>• ____________ |

Name ______________________________

Date ______________________________

# Lesson 1—How Does Water Move from Earth to the Air and Back Again?

## 1. Investigation Skill Practice–Draw Conclusions

**Miguel noticed something on his bus ride to school in the morning. He noticed that when the bus drove through a valley between two hills, it often drove through thick fog. He noticed that the fog was only around on cool mornings, and it was always gone by the afternoon when the sun was shining. What conclusions can you draw from Miguel's observations?**

______________________________________________________________

______________________________________________________________

______________________________________________________________

## 2. Focus Skill Reading Skill Practice–Main Idea and Details

**Read the selection. Underline the main idea. List the details on the lines below.**

Some scientists believe that they can use the sun to convert salt water to fresh water. They know that developing such a process would have advantages and disadvantages. First, the energy to make the fresh water would be cheap, since the sun's energy is free. Second, the source of ocean water is endless. There is more water in the ocean than is needed on land. However, the water conversion equipment would take a lot of money to build. Also, it could be costly to pump the fresh water to the cities or dry farmlands where it is needed.

______________________________________________________________

______________________________________________________________

______________________________________________________________

Name ___________________

## Science Concepts

**3. Answer the Questions**

Why is Earth called the water planet?

___________________

Where is most of the water on Earth?

___________________

How do we get fresh water?

___________________

___________________

**4. On the lines below, write the steps in the water cycle.**

___________________

___________________

___________________

___________________

___________________

___________________

**5. Write *evaporation* or *condensation* to describe what happens in each sentence below.**

Your cold glass becomes wet when it stands outside on a hot day. ___________________

Sweat dries from your face. ___________________

The birdbath you filled last week is now dry and empty. ___________________

Name ______________________________

Date ______________________________

# Lesson 2—How Do Californians Get the Water they Need?

## A. Compound Words

**Write the word or phrase that fits the description below.**

| watershed | groundwater | water table |
|---|---|---|

1. Compound word describing an area that sheds water off its surface ____________
2. Compound word identifying water that is underground ____________
3. Phrase describing the underground table where water pools ____________

## B. Word Origins

**Answer the questions about word origins.**

1. The Greek word *hydor* means "water." How would you define *hydroelectric?*

______________________________________________________________

2. The Latin word *aquae* means "water," and a duct is a type of tube or channel. How would you define *aqueduct?*

______________________________________________________________

3. The Latin word *reservare* means "to hold back." How would you define *reservoir?*

______________________________________________________________

## C. Analogies

**Fill in the blanks with one of the words from the box.**

| dam | aqueduct |
|---|---|

____________ is to *hydroelectric power* as *solar panels* are to *solar power*

____________ is to *water* as *pipeline* is to *natural gas*

Name ________________________

Date ________________________

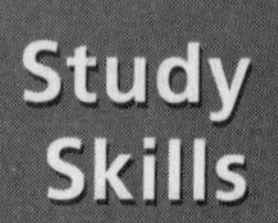




# Lesson 2—How Do Californians Get the Water they Need?

## Understanding Vocabulary

**Using a dictionary can help you learn new words that you find as you read.**

- A dictionary shows all the meanings of a word and tells where the word came from.
- You can use a chart to list and organize unfamiliar words that you look up in a dictionary.

**As you read this lesson, look up unfamiliar words in the dictionary. Add them to the chart below. Fill in each column to help you remember the word's meaning.**

**dam** (DAM) ***n.*** [ME < MDu *dam*, ON *dammr*, <L *facere*] **1** a barrier built to hold back flowing water **2** the water thus kept back
[adapted from *Webster's Third College Edition*]

| Word | Syllables | Origin | Definition |
|---|---|---|---|
| watershed | WAWT•er•shed | compound word. *Water* comes from German | The area of land in which water runs off into a particular system of creeks and rivers |
| Reservoir | | | |
| Aqueduct | | | |
| groundwater | | | |

Name ______________________________

Date ______________________________

# Lesson 2—How Do Californians Get the Water they Need?

## 1. Investigation Skill Practice–Plan and Conduct an Investigation

Nancy was worried that she used too much water. She planned an experiment to find out how much water she could save by investigating with a simple water-saving plan. First, she measured how much water came from the tap while she brushed her teeth. Then she brushed her teeth using a cup of water instead. She measured how much water she saved. She calculated how much water she could save if she did the same thing every day. That's a lot of water!

What was Nancy's plan for her investigation?

______________________________

______________________________

How did she conduct her investigation?

______________________________

______________________________

______________________________

## 2. Focus Skill Reading Skill Practice–Main Idea and Details

**Read the selection. Underline the main idea. Write 3 details on the lines below.**

Families who draw their water from a well take extra care to conserve water outside and inside their homes. They often plant native species of plants so that they do not need to water their lawn. They collect rain water to feed their animals and wash their cars. They also use water wisely by taking short showers, turning off water when it is not being used, and by washing only full loads of laundry.

______________________________

______________________________

Name ______________________________

## Science Concepts

**3. Fill out the chart below by explaining if the type of water supply to the left is made by humans or nature and how it helps supply water.**

| How Do Californians Get the Water they Need? | | |
|---|---|---|
| | **Made by Humans or Nature?** | **How does it help supply water to Californians?** |
| lakes and rivers | | |
| watershed | | |
| aqueduct | | |
| reservoirs | | |
| groundwater | | |
| aquifers | | |
| wells | | |

**4. On the lines below, explain why saltwater is a valuable resource to Californians.**

______________________________

______________________________

Name ______________________

Date ______________________

# Lesson 3—How Can People Conserve Water?

## A. The Suffix *–tion*

The suffix *–tion* changes the meaning of the base word to which it is added. The suffix *–tion* means, "the act of [doing something]" or "the thing that is [being done]" or "the state of being [a certain way]."

**Read each base word in the table. Choose the word from the box that is the base word plus the suffix *–tion*, and write it in the table. Then write its meaning. You can use a glossary to help you.**

| pollution | conservation | reclamation | irrigation |
|---|---|---|---|

| Base Word | New Word | Meaning |
|---|---|---|
| 1. irrigate | | |
| 2. pollute | | |
| 3. conserve | | |
| 4. reclaim | | |

## B. Explore Word Meanings

**Circle the word that answers the question.**

1. What phrase describes the safety of water for use by humans, animals, and plants?

   water table    water quality    watershed

2. What is the word for a period of little rain?

   summer    desert    drought

Name ______________________

Date ______________________

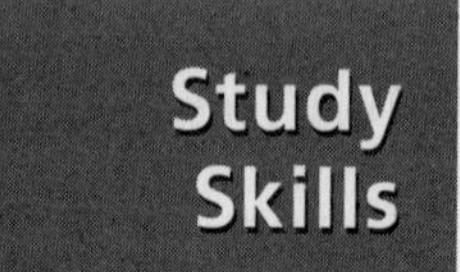




# Lesson 3—How Can People Conserve Water?

## Write to Learn

**Writing about what you read can help you better understand and remember information.**

- Writing down the information that you learn from each lesson leads you to think about the information.
- Writing your own response to the new information makes it more meaningful to you.

**As you read the lesson, pay attention to new and important information. Keep track of the information and your responses in the log below.**

| How Can People Conserve Water? | |
|---|---|
| **What I Learned** | **Personal Response** |
| Whatever I pour down the drain will end up in a body of water. | Now I understand how water can get polluted by people who do not think carefully about what they put down the drain. |
| | |
| | |
| | |
| | |
| | |

Name ______________________________

Date ______________________________

# Lesson 3—How Can People Conserve Water?

## 1. Investigation Skill Practice–Identify a Single Independent Variable

Julio planned an experiment where he compared the water used by two homeowners. Both homeowners had similar amounts of property. They lived in the same area of town, so they experienced the same amount of rainfall. However one homeowner had a Xeriscape that included a rock garden, native grasses, and large bushes in his yard. The second homeowner planted a closely trimmed lawn of grass that needed water each week. Julio plans to compare the water bills for both homeowners to find out who uses more water.

What do you think Julio plans to prove by his experiment?

______________________________________________________________

What is the independent variable in Julio's experiment?

______________________________________________________________

## 2. Focus Skill Reading Focus Skill Practice–Cause and Effect

**Read the selection. Describe the cause and effect of Cheryl's actions.**

Cheryl walked to the river one day and saw that it had garbage in it. She smelled a bad odor coming from the water, and she noticed that there were no birds or fish near the water. She decided to try to get her town to clean up the river. Cheryl went to the town council meeting to ask them to pass a law about dumping in the river. She organized clean-up days to collect trash from the river. Finally, she wrote letters to the factories along the river to ask them to protect the water in the town. All her efforts paid off. Two years later, the river was clean. It did not smell bad, and there were fish and ducks swimming in it all day.

______________________________________________________________

______________________________________________________________

______________________________________________________________

Name ________________________

Science Concepts

**3. Circle the substances that can pollute water.**

| | | |
|---|---|---|
| fertilizer | toothpaste | dumped material from mines |
| food coloring | sewage | |
| | | milk |
| pesticides | factory waste | |

**4. Methods of Conserving Water**

**Explain how each method below helps conserve water resources.**

water treatment by factories

________________________

________________________

water reclamation

________________________

________________________

Xeriscaping

________________________

________________________

drip irrigation

________________________

________________________

Name ______________________

Date ______________________

# Lesson 3—How Can People Conserve Water?

## A. Daily Water Usage

**We all use water in our daily living. Below is a chart which shows some of the ways we use water and the amounts that we use.**

| Water Used For | Gallons Used | Water Used For | Gallons Used |
|---|---|---|---|
| bathroom sink | 1 per use | kitchen sink | 3 per use |
| clothes washer | 30 per load | shower | 40 per use |
| dishwasher | 12 per load | toilet | 2 per flush |

**1.** Which thing uses the greatest amount of water each time it is used?

______________________________________________

**2.** Arrange the items in the above list from least gallons used each time to the most gallons used each time.

______________________________________________

**3.** Assume you use the toilet 6 times and the bathroom sink 6 times and the shower one time. Complete the table below to find out how many gallons of water a day you might use.

| Water Used For | Number of Times | Total Gallons Used |
|---|---|---|
| bathroom sink | | |
| shower | | |
| toilet | | |
| Total Used | XXXXXX | |

**4.** Assume your family does 5 loads of wash in a week and runs the dishwasher two times a day. How many gallons of water is used in one week?

______________________________________________

______________________________________________

______________________________________________

Name

## B. Dams

**There are many dams that have been built to meet the needs of the people of California. Below is a chart showing information about seven of these dams.**

| Dam | Year Built | Capacity (acre-ft) |
|---|---|---|
| Barrett | 1922 | 44,755 |
| Calaveras | 1925 | 100,000 |
| El Capitan | 1934 | 112,800 |
| Lake Almanor | 1927 | 1,308,000 |
| Nacimiento | 1957 | 350,000 |
| Oroville | 1968 | 3,537,577 |
| San Antonio | 1965 | 350,000 |

**1.** What functions do dams in California serve?

**2.** Rearrange the chart to show the dams from greatest capacity to least capacity.

| Dam | Year Built | Capacity (acre-ft) |
|---|---|---|
| | | |
| | | |
| | | |
| | | |
| | | |
| | | |
| | | |

**3.** What is the total storage capacity of these seven dams?

Name ______________________

Date ______________________

# Lesson 1—How Does Uneven Heating of Earth Affect Weather?

## A. Explore Word Meanings

**Match the clue on the left to the term on the right.**

1. _____ weather
2. _____ atmosphere
3. _____ air pressure
4. _____ convection current
5. _____ prevailing wind
6. _____ troposphere

a. The blanket of air surrounding Earth

b. The layer of air closest to Earth's surface

c. The condition of the atmosphere at a certain place and time

d. The weight of the atmosphere pressing down on Earth

e. The upward and downward movement of air in the atmosphere

f. A global wind that blows constantly from the same direction

## B. Context Clues

**Complete the sentence with the correct word from the box.**

| air pressure | local wind | stratosphere | prevailing winds |
|---|---|---|---|

1. The lower ______________ at high altitudes means that the air is thinner and more difficult to breathe.
2. We noticed that the ______________ around our lake created a pleasant breeze even on the hottest days.
3. We worry about how our actions on Earth can affect the ozone in the ______________.
4. The ______________ at the equator allowed traders to easily sail across the ocean.

Name ______________________________

Date ______________________________

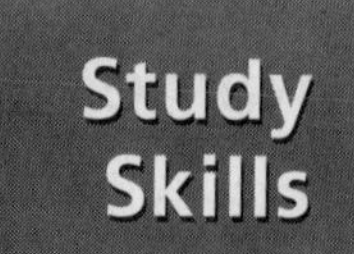




# Lesson 1—How Does Uneven Heating of Earth Affect Weather?

## Preview and Question

**Identifying main ideas in a lesson and asking questions about them can help you find important information.**

- To preview a lesson, read the lesson title and the section titles. Look at the pictures, and read their captions. Try to get an idea of the main topic and think of questions you have about the topic.
- Read to find the answers to your questions. Then recite, or say, the answers aloud. Finally, review what you have read.

**As your read this lesson, fill in the chart and practice reading, reciting, and reviewing.**

| How Does Uneven Heating of Earth Affect Weather? | | | | |
|---|---|---|---|---|
| **Preview** | **Questions** | **Read** | **Recite** | **Review** |
| The Atmosphere: Weather is the condition of the atmosphere at a certain place and time. | How does the atmosphere change from one place to another? | ✔ | ✔ | ✔ |
| ______________________ ______________________ ______________________ ______________________ ______________________ | ______________________ ______________________ ______________________ ______________________ ______________________ | | | |

Name ______________________

Date ______________________

# Lesson 1—How Does Uneven Heating of Earth Affect Weather?

1. **Investigation Skill Practice–Predict**

Tony decided to ride his bike from his home in New York all the way across the country to Los Angeles. He packed lots of gear and made plans to make plenty of stops. He practiced riding for long distances before he started. But one thing he did not account for was the prevailing westerlies. Those are the winds that usually blow from west to east. How do you predict that the wind will affect his trip from New York to Los Angeles?

_______________________________________________

_______________________________________________

_______________________________________________

2. Focus Skill **Reading Focus Skill Practice–Main Idea and Details**

**Read the selection. Underline the main idea. Write two details on the lines below.**

No one really knows how the city of Chicago got the nickname "The Windy City." Some people think that the city got the name because it is on the edge of Lake Michigan. Wind from the lake blows toward the land because the air over the city heats up and rises. Others believe that the nickname came from Chicago's politicians who are famous for talking so loudly and for so long.

_______________________________________________

_______________________________________________

_______________________________________________

_______________________________________________

Name ______________________

## Science Concepts

**3. Circle the phrase that completes each sentence.**

**a.** Air pressure is greatest ____________.

at sea level — at the top of a mountain

**b.** Most of the Earth's weather occurs in the ____________.

troposphere — stratosphere

**c.** Air pressure pushes air ____________.

downward — in all directions

**d.** ____________ air is heavier.

Cold air — Warm air

**e.** The sand is warmer than the water on a sunny day because ____________.

water heats up faster than land — land heats up faster than water

**f.** A sea breeze occurs when ____________.

the wind blows from the sea to the land — the wind blows from the land to the sea

**g.** The prevailing wind in the United States blows ____________.

from west to east — from east to west

**h.** The sun's rays are more direct ____________.

in the polar regions — at the equator

**i.** Weather in the United States generally moves ____________.

from California toward the east — from New York toward the west

**j.** The wind around the equator is called the ____________.

prevailing westerly — trade winds

Name ______________________

Date ______________________

# Lesson 2—How Do the Oceans and the Water Cycle Affect Weather?

## A. Multiple Meaning Words

**Some words can have different meanings depending on the context of the sentence in which they are used. Read the sentence. Circle the correct meaning of the underlined word.**

**1.** As the ships sailed west, the ocean current helped push them along their way.

stream of water                    taking place in the present day

**2.** When the water on the surface of the ocean is warm, it evaporates quicker.

the uppermost layer                    to come up for air

**3.** The coast of California is affected by ocean water that brings cool weather.

to glide without effort                    the area of land near the ocean

**4.** The warm water in the Gulf Stream brings warm temperatures to northern Europe.

a wide impassible gap                    the origin of a major ocean current

## B. Parts of Speech

**Choose the correct form of the word to complete the sentence.**

**1.** humidity/humid

**Noun:** The scientist measured the ______________ by finding the amount of water vapor in the air.

**Adjective:** She found that the air today is very ______________.

**2.** precipitation/ precipitate

**Noun:** The forecaster predicts some form of ______________ for today.

**Verb:** We wondered if the clouds would ______________ during our picnic.

Name ____________________

Date ____________________

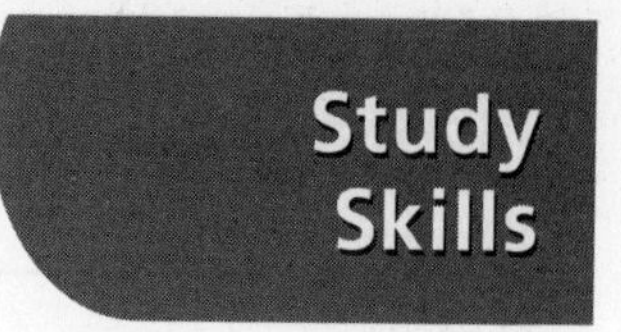




# Lesson 2—How Do the Oceans and the Water Cycle Affect Weather?

## Skim and Scan

**Skimming and scanning are two ways to learn from what you read.**

- To skim, quickly read the lesson title and the section titles. Look at the visuals, or images, and read the captions. Use this information to identify the main topics.
- To scan, look quickly through the text for specific details, such as key words or facts.

**Before you read this lesson, skim the text to find the main ideas. Then look for key words. If you have questions about a topic, scan the text to find the answers. Fill in the chart below as you skim and scan.**

| How Do the Oceans and the Water Cycle Affect Weather? | |
|---|---|
| **Skim** | **Scan** |
| Lesson Title: | Key Words and Facts: |
| Main Idea: | |
| Section Titles: | |
| Visuals: | |

Name ______________________

Date ______________________

# Lesson 2—How Do the Oceans and the Water Cycle Affect Weather?

1. **Investigation Skill Practice–Infer**

**Some factories release steam and hot gasses into the air. Such warm air rises into the atmosphere and warms the colder layers. Based on your knowledge of weather patterns, what can you infer about the effect of hot gasses on the weather of a region?**

______________________________________________

______________________________________________

______________________________________________

2. Focus Skill **Reading Focus Skill Practice–Cause and Effect**

**Read the selection. Describe the cause and effect of the air mass over Siberia.**

The Siberian High is a large air mass that stays over Siberia in winter. The air mass forms as snow on the ground, and it cools air over the land. Siberia is so far north that it lacks enough sunlight to melt the snow or to warm the air. The air mass causes great air pressure and cold weather in Siberia and other parts of Asia and Europe.

______________________________________________

______________________________________________

______________________________________________

______________________________________________

Name ______________________________

## Science Concepts

**3. Put the steps in each cycle in the correct sequence. The first step is already completed for you.**

### A. The Water Cycle

____ Water vapor condenses into cloud drops.

____ Water soaks into the ground or falls into streams, rivers, lakes, and the oceans.

__1__ Water evaporates from the oceans.

____ Water falls to Earth as rain, sleet, snow, or hail.

### B. Global Winds

____ Cool northern regions receive warm weather when the heat from the tropics is released.

__1__ Water evaporates from the warm water of the tropics.

____ Global winds push warm air masses away from the topics.

### C. Clouds

____ The condensed water forms a cloud.

__1__ Warm air rises.

____ Water begins to condense as the air cools.

____ As it rises, the warm air cools.

### D. Frost

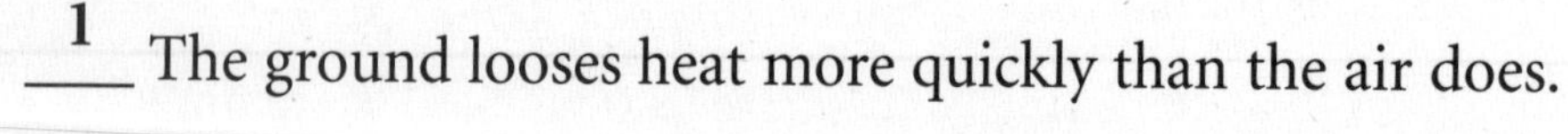

__1__ The ground looses heat more quickly than the air does.

____ If the air is cold enough, frost forms.

____ As the ground cools, water vapor condenses near the ground.

Name ______________________________

Date ______________________________

# Lesson 3—How Is Weather Predicted?

## A. Graphic Organizer

**Fill in the blanks with the correct word from the box and definition.**

| barometer | anemometer | hygrometer |
|---|---|---|

| Word | Definition | What does it show? |
|---|---|---|
| 1. | | shows that a low pressure system is coming |
| 2. | | shows that the wind is blowing at 15 miles per hour |
| 3. | | shows that the humidity is 85% |

## B. Context Clues

**Complete the sentence with one of the words from the box.**

| meteorology | meteorologist | front | air mass | forecasted |
|---|---|---|---|---|

We watched the ________________ on television to find out what the weather will be like tomorrow. She ________________ that there will be a large, warm ________________ coming toward our area. It will bring humidity and warm days. However, before the warm days are here, the warm air will run into the cold air causing a ________________. Thanks to the science of ________________ we know what the weather will be tomorrow.

Name ______________________

Date ______________________

**Study Skills**

Unit 4, Lesson 3

# Lesson 3—How Is Weather Predicted?

## Connect Ideas

**You can use a web organizer to show how different ideas and information are related.**

- List important themes in the ovals in the web's center.
- Add ovals showing main ideas that support each theme.
- Add bubbles for the details that support each main idea.

**Complete the web as you read this lesson. Fill in each bubble by adding facts and details. Add your own bubbles as you need them.**

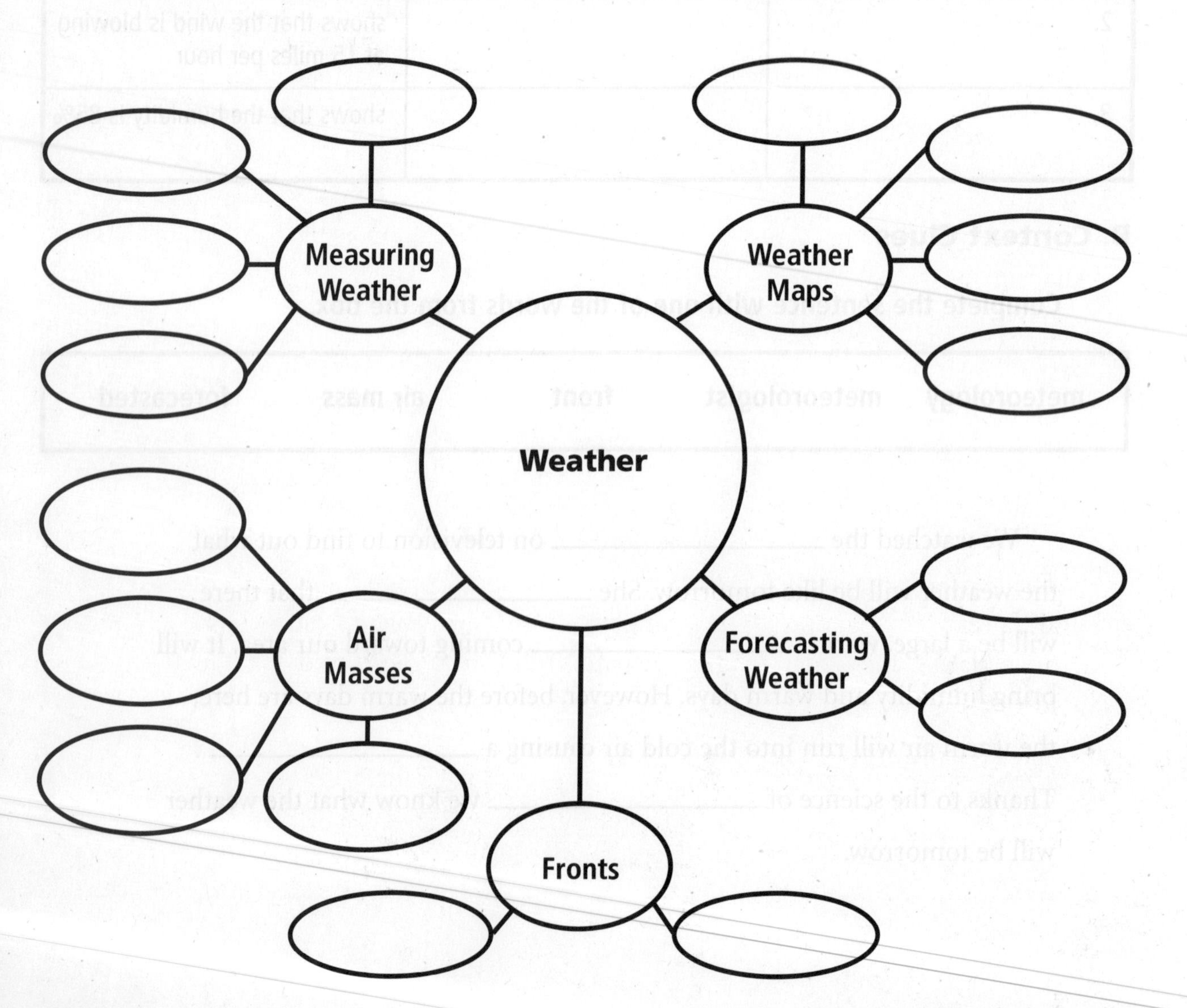

Name ______________________________

Date ______________________________

# Lesson 3—How Is Weather Predicted?

**1. Investigation Skill Practice–Gather and Record Data**

Robert wanted to be a meteorologist. He began by watching the weather in his own region. As he watched the weather and recorded data, he began to notice patterns. He looked for relationships between wind direction and rain, between air pressure and rainfall, and between temperature and air pressure. He even tried to explain what was happening when a weather pattern seemed unusual.

What records do you think Robert kept that helped him make his observations?

______________________________

______________________________

How is gathering and recording data important to learning about weather patterns?

______________________________

______________________________

______________________________

**2. Focus Skill Reading Skill Practice–Main Idea and Details**

**Read the selection. Underline the main idea. Write 3 details on the lines below.**

Santa Ana winds are famous for their part in strengthening forest fires. The winds move through the desert and over and through mountains. The air is very warm and dry. The strong winds can turn small forest fires into a huge blaze. Often everything in the path of the fires is destroyed, including forests, homes, and businesses.

______________________________

______________________________

______________________________

Name ______________________

Science Concepts

3. **For each scenario, tell what kind of weather might be coming your way.**

You notice that the wind in your area has begun to blow from the north.

______________________

You read your barometer and notice that the air pressure is dropping.

______________________

You see large, gray cumulus clouds coming your way.

______________________

The reading on your hygrometer shows a rapid increase in humidity.

______________________

The weather map shows that the area west of you has snow storms and the wind direction is blowing east.

______________________

A cold front is coming quickly to your area where you now have warm humid weather.

______________________

______________________

Today you have cold weather but a warm front is coming.

______________________

______________________

Your area is getting a continental polar air mass.

______________________

A maritime tropical air mass is predicted for your region.

______________________

Name ______________________

Date ______________________

**Extra Support**

Unit 4, Lesson 3

# Lesson 3—How Is Weather Predicted?

## A. Reading Weather Maps

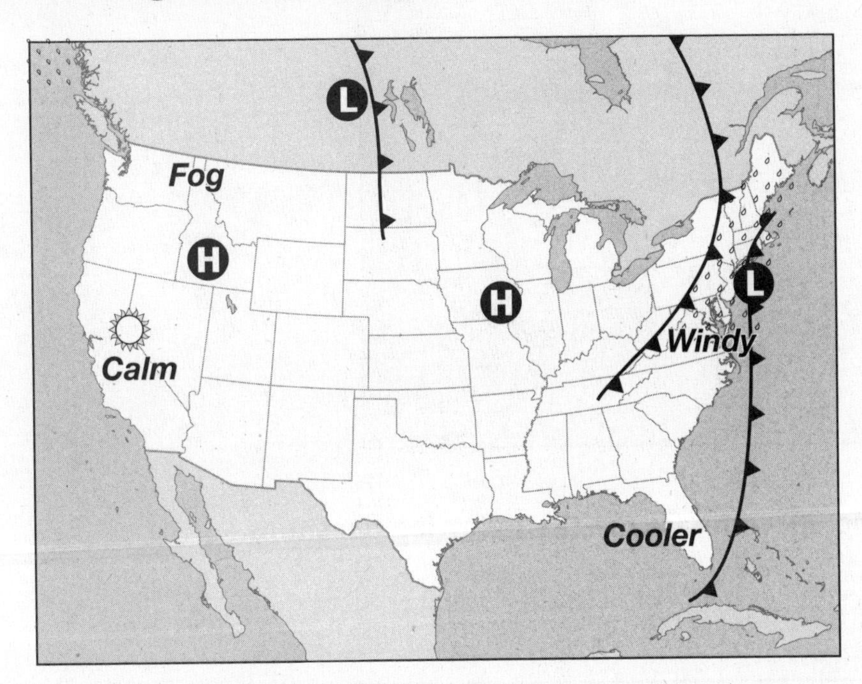


**Use the weather map shown above to answer the following questions.**

**1.** What is the weather like in California?

______________________

**2.** Which part of the United States is experiencing windy conditions?

______________________

**3.** What does the letter "H" stand for on the map?

______________________

**4.** Name one state that is experiencing fog according to the map.

______________________

Name ______________________________



**Use the weather map above to answer the following questions.**

**1.** What type of pressure system is in Ohio?

______________________________________________

**2.** What type of front is moving through Wisconsin and northern Illinois?

______________________________________________

**3.** Describe which part of the country a warm front is moving through.

______________________________________________

**4.** Describe which part of the country a cold front is moving through.

______________________________________________

Name ______________________________

## B. Creating Weather Maps

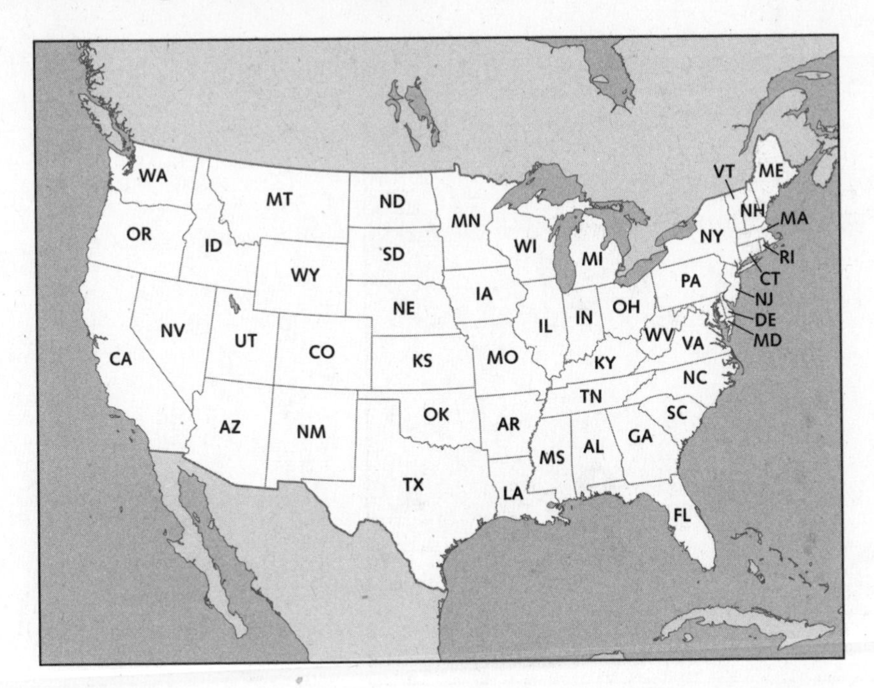


**Use the United States map shown above to complete the exercises described below.**

**1.** Create a weather map using the map above with the following characteristics:

- High pressure systems in the Utah/Nevada area and in the New York area.
- Low pressure system in Iowa
- A cold front that goes from Ohio to Georgia
- A warm front in the California/Nevada area

Name ______________________________

## C. Interpreting Weather Maps

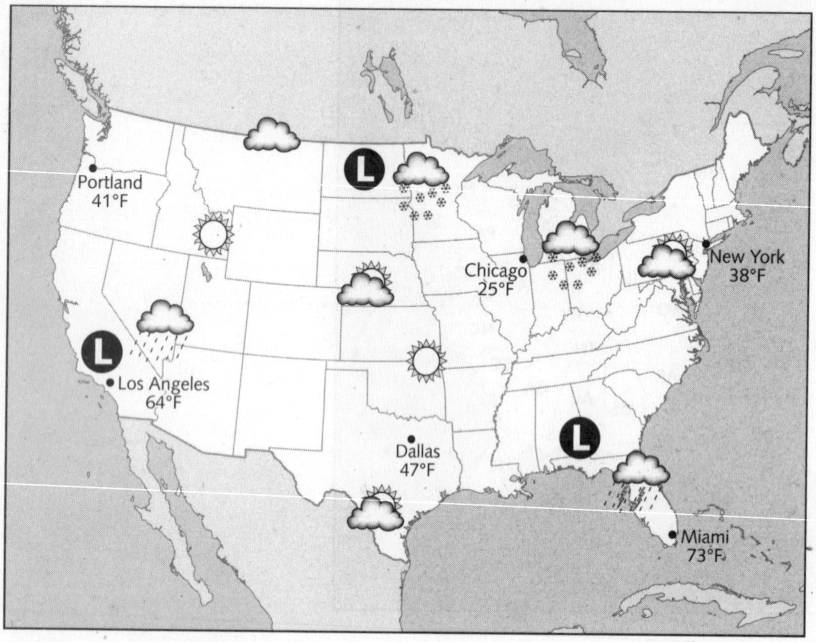

**Use the weather map above to answer the following questions.**

**1.** Name a state that is receiving rain.

______________________________

**2.** What is the weather like in Dallas, Texas?

______________________________

**3.** What is the weather like in northwestern Ohio and Northeastern Indiana?

______________________________

**4.** Name one part of the country that has sunny weather.

______________________________

**5.** What type of weather is usually experienced in a high-pressure area?

______________________________

**6.** What type of weather is usually experienced in a low-pressure area?

______________________________

Name ______________________

Date ______________________

# Lesson 4—What Are the Causes and Effects of Severe Weather?

## A. Work Origin/Etymology

**Choose one word from the box that is derived from the word given. Write the word and its definition in the table.**

| monsoon | hurricane | thunderstorm | tornado |
|---|---|---|---|

| Word | Derived From | Language of Origin | Definition |
|---|---|---|---|
| | *huracan* | Spanish | |
| | *mawsim* | Arabic | |
| | *tronada* | Spanish | |
| | *thunor* | Old English | |

## B. Write a sentence for each word listed.

monsoon

______________________

hurricane

______________________

tornado

______________________

Name ______________________

Date ______________________

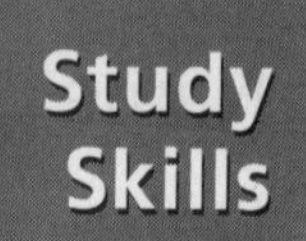




# Lesson 4—What Are the Causes and Effects of Severe Weather?

## Write to Learn

**Writing about what you read can help you better understand and remember information.**

- Writing down the information that you learn from each lesson leads you to think about the information.
- Writing your own response to the new information makes it more meaningful to you.

**As you read the lesson, pay attention to new and important information. Keep track of the information and your responses in the log below.**

| What Are the Causes and Effects of Severe Weather? | |
|---|---|
| **What I Learned** | **Personal Response** |
| The monsoons are a seasonal event caused by a shift in wind direction. | Wet, humid, rainy days flow one to another endlessly, until they end and go away as if they had never been here at all. |
| | |

Name ______________________________

Date ______________________________

# Lesson 4—What Are the Causes and Effects of Severe Weather?

## 1. Investigation Skill Practice–Predict

When you hear the National Weather Service announce that a hurricane is on the way for your area, you must make certain preparations. Based on what you know about severe weather, what can you predict will happen when the hurricane arrives?

______________________________________________________________

______________________________________________________________

How would your prediction of the conditions be different if the weather service had announced a thunderstorm for your area?

______________________________________________________________

______________________________________________________________

## 2. Focus Skill Reading Skill Practice–Cause and Effect

**Read the selection. Describe the cause and effect of a storm surge.**

A storm surge is a flow of ocean water that rises during a hurricane. It is caused by strong winds that push water ahead of the storm. A storm surge can cause more damage than the hurricane's high winds or rain. That's because water is extremely heavy. It moves with great power and can smash buildings and vehicles. It can push houses over, knock trees down, and carry cars away. In addition, the water often carries heavy debris. When the storm surge recedes, it leaves mountains of branches, broken equipment, and dirt. Such debris is difficult and expensive to remove.

______________________________________________________________

______________________________________________________________

______________________________________________________________

______________________________________________________________

______________________________________________________________

Name ______________________

## Science Concepts

**3. Fill in the chart with accurate information about each type of storm.**

| | tornado | thunderstorm | monsoon | hurricane | typhoon |
|---|---|---|---|---|---|
| duration of storm | | | | | |
| size of storm | | | | | |
| cause of storm damage | | | | | |
| causes of storm | | | | | |
| predictability | | | | | |

**4. Answer the questions.**

What is the difference between a hurricane and a typhoon?

______________________

______________________

What is the eye of a hurricane?

______________________

______________________

Name ______________________

Date ______________________

# Lesson 1—What Is the Sun?

## A. Analogies

**An analogy is made of two pairs of words. The words in each pair are related in the same way. Think about the relationships of the following pairs of words. Then choose a word from the box to complete the analogy.**

| star |
|---|
| sun |
| fusion |

**1.** *Sun* is to *day* as ______________ is to *night.*

**2.** ______________ is to *star* as *wood* is to *fire.*

**3.** *Water* is to *hydroelectric dam* as ______________ is to *solar panel.*

## B. Categorizing and Classifying

**In each group below, one word does not belong to the same category as the others. Circle the letter of that word. Then classify the remaining words by writing a label for that group.**

**1.** A. sun
B. moon
C. planets
D. fusion

______________________________

**2.** A. color
B. number of planets
C. distance from Earth
D. size

______________________________

**3.** A. X rays
B. visible light
C. fusion
D. radio waves

______________________________

Name ______________________________

Date ______________________________

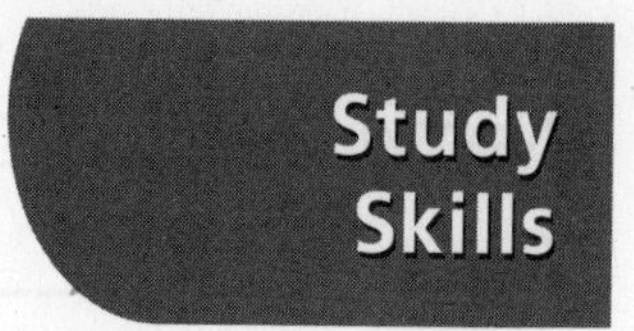




# Lesson 1—What Is the Sun?

## Anticipation Guide

**An anticipation guide can help you anticipate, or predict, what you will learn as you read.**

- Look at the section titles for clues.
- Preview the Reading Check question at the end of each section. Use what you know about the subject of each section to predict the answers.
- Read to find out whether your predictions were correct.

**As you read the section, complete your own anticipation guide below. Predict answers to each question and check to see if your predictions were correct.**

| What Is the Sun? | | |
|---|---|---|
| **Stars** | | |
| **Reading Check** | **Prediction** | **Correct?** |
| What are two ways that scientists classify stars? | In this section, we will learn about different types of stars. | |
| **Features of the Sun** | | |
| **Reading Check** | **Prediction** | **Correct?** |
| What are the layers of the sun? | | |
| **How the Sun Produces Energy** | | |
| **Reading Check** | **Prediction** | **Correct?** |
| How does the sun produce energy? | | |

Name ______________________

Date ______________________

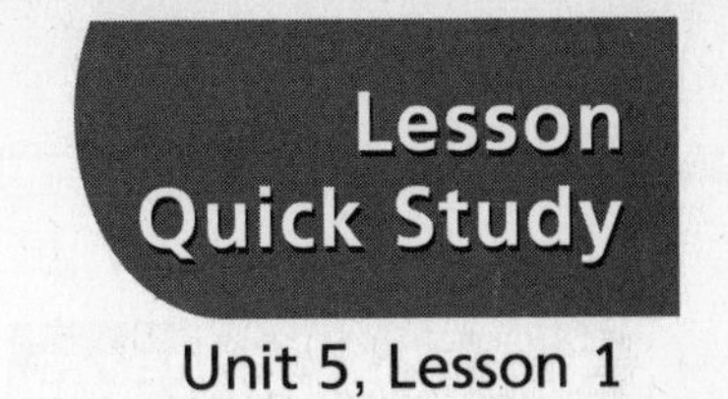




# Lesson 1—What Is the Sun?

## 1. Investigation Skill Practice–Predict

Solar power relies on sunny days, long summers, and short winters. Panels collect the sun's energy and transformers change it into electricity. However, solar energy is not practical all the time. When the sun does not shine, the system cannot collect the sun's energy. In addition, solar panels and the equipment it requires are expensive to install. It is difficult to install a solar heating system in a house that has already been built.

**Make predictions to answer the following questions.**

Where do you think solar panel systems might sell the best and the worst?

______________________________________________

______________________________________________

Do you predict that solar heating will become popular throughout the nation? Tell why or why not.

______________________________________________

______________________________________________

## 2. Focus Skill Reading Skill Practice–Main Idea and Details

**Read the selection. Underline the main idea. Write two details on the lines below.**

Icarus and Daedalus are mythical characters who are known for their unsuccessful flight toward the sun. Daedalus built wings for himself and his son so they could leave the island of Crete. He fastened feathers to the wings with wax. Then the two flew away from the island. However, his son, Icarus, flew too close to the sun. The sun's warmth melted the wax on his wings, and he fell to his death in the sea.

______________________________________________

______________________________________________

______________________________________________

Name ______________________________

## Science Concepts

**3. Match the description with one of the words from the box.**

| |
|---|
| **photosphere** |
| **core** |
| **radiation zone** |
| **convection zone** |

**a.** where sunspots appear ____________

**b.** produces solar flares ____________

**c.** sits at the center of the sun ____________

**d.** contains most of the sun's mass ____________

**e.** energy passes through zone toward surface ____________

**f.** the area that can be seen on Earth ____________

**g.** sun's outer zone ____________

**4. Answer the questions about the sun.**

Is it the brightest star? ____________

Is it the hottest star? ____________

Is it the biggest star? ____________

Is it the largest object in the solar system? ____________

Is it the closest star to Earth? ____________

**5. Answer the questions about fusion.**

What happens during fusion?

______________________________

______________________________

What kinds of waves come from the sun?

______________________________

______________________________

Name ______________________________

Date ______________________________

# Lesson 2—What Makes Up the Solar System?

## A. Context Clues

**Complete the sentence with the correct word from the box.**

| comet<br>meteors<br>planet<br>satellite<br>solar system |
|---|

Ida dreams of traveling through space and seeing all the objects of the ______________. She hopes to see the ______________ including Jupiter, Saturn, and Venus. She is interested in Jupiter because of its many ______________. She wonders what it would be like to live in a place with so many moons. Ida hopes that her dream of space travel would come true someday. In the meantime, she observes the sky on her own. She looks for ______________ that appear at intervals of many years. She tracks ______________ that enter the earth's atmosphere as shooting stars. She also studies hard at school to learn all she can about science and space.

## B. Matching

**Write the letter of the definition from the right next the correct word on the left.**

1. ______ comet
2. ______ solar system
3. ______ asteroid
4. ______ satellite
5. ______ planet

**a.** a star and all the planets and other objects that revolve around it

**b.** a body that revolves around a star

**c.** a body in space that orbits a larger body

**d.** a piece of rock and metal that orbits the sun, forming a belt between Mars and Jupiter

**e.** a ball of ice, rock, and frozen gases that orbits the sun

Name ______________________

Date ______________________

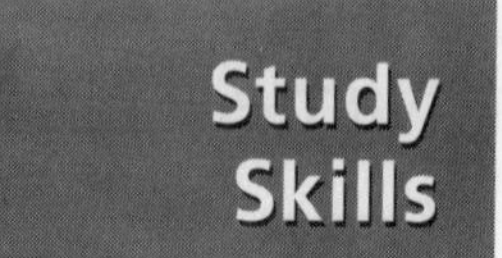




# Lesson 2—What Makes Up the Solar System?

## Pose Questions

**Posing, or asking questions, as you read can help you understand what you are reading.**

- Form questions as you read. For example, you may ask how a science concept is connected to other concepts.
- Use the questions to guide your reading. Look for answers as you read.

**Before you read this lesson, write a list of questions in the chart below. Look for the answers as you read. Record the answers in the chart.**

| What Makes Up the Solar System? | |
|---|---|
| **Questions** | **Answers** |
| What separates the inner planets from the outer planets? | The asteroid belt between Mars and Jupiter divides the inner planets from the outer planets. |
| | |
| | |
| | |

Name ______________________________

Date ______________________________

# Lesson 2—What Makes Up the Solar System?

## 1. Investigation Skill Practice–Infer

Many of the brightest stars and galaxies in the sky have Arabic names, such as Rigel, Aldebaran, Vega, Deneb, Shedir, and Betelgeuse. They were named very long ago. However, some stars that are not very bright are unnamed. They are designated only by their telescopic coordinates.

**Make inferences to explain why many bright stars have Arabic names.**

**Make inferences to explain why dim stars have no names.**

## 2. Focus Skill Reading Skill Practice–Main Idea and Details

**Read the selection. Underline the main idea. Write three details on the lines below.**

Edmond Halley first figured out that a streak in the sky was a comet that moves around the sun every 76 years. He did this by watching the sky and by doing research about comets throughout history. He found out that every 76 years people reported a bright streak in the sky. No one could explain it, but Halley determined that these reports must have referred to the same comet. His research and attention to history helped him discover the truth about a mysterious comet that bears his name today.

______________________________

______________________________

______________________________

______________________________

______________________________

Name ______________________________

## Science Concepts

**3. Answer the questions below about the planets.**

What are the planets of the solar system?

______________________________

Which planets are gas giants?

______________________________

Which are the inner planets?

______________________________

What divides the planets into inner planets and outer planets?

______________________________

Why do some people argue that Pluto is not really a planet?

______________________________

**4. Answer the questions below about asteroids.**

How big is an asteroid?

______________________________

Where are most asteroids found?

______________________________

**5. Answer the questions below about comets.**

What does a comet look like?

______________________________

When can you see a comet?

______________________________

How is a comet different from a star?

______________________________

How is a comet different from a planet?

______________________________

Name ______________________

Date ______________________

# Lesson 2—What Makes Up the Solar System?

## A. Space Research/Exploration—Internet Research

**Visit the NASA government website at *http://www.nasa.gov***
**Click on the *Missions* button. Select one of the *Current Missions* listed on the website. Click on the icon for that mission and read about it. Use the information you read to answer the questions below.**

**1.** What is the name of the mission?

______________________________________________

**2.** When does the mission begin and end?

______________________________________________

______________________________________________

**3.** Describe the purpose of the mission.

______________________________________________

______________________________________________

______________________________________________

______________________________________________

______________________________________________

**4.** Is the mission manned or unmanned?

______________________________________________

**5.** What parts of the solar system will be explored during this mission?

______________________________________________

______________________________________________

______________________________________________

Name ______________________

## B. Space Research/Exploration—Research Halley's Comet

**Use the library or internet to research Halley's Comet. Answer the questions below.**

**1.** What causes the tail on Halley's Comet?

______________________

**2.** Why does Halley's Comet travel through our solar system regularly?

______________________

**3.** Which direction does a comet's tail point?

______________________

**4.** When is the next time we will be able to see Halley's Comet?

______________________

**5.** Who is the comet named after?

______________________

**6.** Write a short paragraph explaining how people have observed Halley's Comet since 1682?

______________________

______________________

______________________

______________________

______________________

______________________

**7.** The center of Halley's Comet is called the nucleus. Write three sentences explaining this comet's nucleus.

______________________

______________________

Name ________________________________

Date ________________________________

# Lesson 3—What Holds the Moon and Planets in Place?

## A. Context Clues

**Read the sentences. Use the context clues to help you choose the correct meaning of the underlined word. Circle its meaning. Look up the word in a glossary if you need help.**

**1.** The orbit of Pluto is unusual because it crosses Neptune's orbit, making it come closer to the sun during some periods and farther away in other periods.

the path something takes as it revolves around another body

oval-shaped

the attraction between all objects

**2.** The planets' path around the sun is not circular, it is elliptical.

long

inconsistent

oval-shaped

**3.** The car's inertia as it skidded on wet ice made it difficult to stop.

slippery tires

tendency to keep moving

bad breaks

**4.** When you throw a ball in the air, Earth's gravity brings the ball back to the ground.

attraction between all objects in the universe

orbit

oxygen content

**5.** The star was so small that it could only be seen through a telescope.

an instrument to see very small things

an instrument to measure air pressure

an instrument to see things far away

Name ______________________________

Date ______________________________

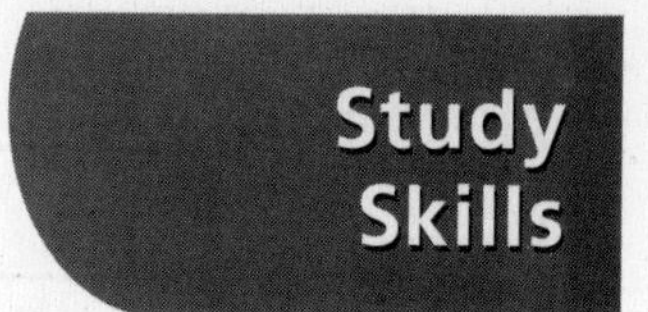




# Lesson 3—What Holds the Moon and Planets in Place?

## Take Notes

**Taking notes can help you remember important ideas.**

- Write down important facts and ideas. Use your own words. You do not have to write in complete sentences.
- One way to organize notes is in a chart. Write down the main ideas in one column and facts and details in another.

**As you read this lesson, use the chart below to take notes.**

| What Holds the Moon and Planets in Place? | |
|---|---|
| **Main Ideas** | **Facts** |
| • Ptolemy believed the sun revolved around the Earth. | • Ptolemy's model was wrong. |
| | • ______________________ |
| • ______________________ | • ______________________ |
| • ______________________ | • ______________________ |
| • ______________________ | • ______________________ |
| • ______________________ | • ______________________ |
| • ______________________ | • ______________________ |
| • ______________________ | • ______________________ |

Name ______________________

Date ______________________

# Lesson 3—What Holds the Moon and Planets in Place?

1. **Investigation Skill Practice–Model**

**Glen made a model to help him understand the phases of the moon. He sat in a dark room with a flashlight and a white softball. He put the flashlight on a table and shined it on the ball. Then he held the ball making the light reflect from the surface. He held the ball so that he could see one entire side of the ball lit up. Then he moved it again so that he could see only half of the ball lit up. Finally, he held the ball so that the ball appeared to be completely in shadow except for a tiny sliver of light on one side.**

What did the flashlight represent in Glen's model?

______________________________________________

What did the ball represent?

______________________________________________

How did his model help explain the phases of the moon?

______________________________________________

______________________________________________

2. Focus Skill **Reading Focus Skill Practice–Cause and Effect**

**Read the selection. Describe the cause and effect relationships.**

All objects with mass in the solar system have their own gravitational pull. For example, the gravity of the sun pulls on the Earth. The gravity of Earth pulls on the moon. The moon also has its own gravitational force. We can tell that this is true because the gravity of the moon pulls on the ocean water on Earth. This pull toward the moon is what creates high tide and low tide all over the world.

______________________________________________

______________________________________________

Name ______________________________

## Science Concepts

**3. A myth is not based on truth. Explain why these myths are untrue.**

Myth 1: The moon, sun, and planets revolve around the Earth.

______________________________

______________________________

______________________________

Myth 2: The planets move in circular orbits around the sun.

______________________________

______________________________

______________________________

Myth 3: The Earth will one day fall into the sun.

______________________________

______________________________

______________________________

Myth 4: The moon changes shape every night, which is why it always looks different.

______________________________

______________________________

______________________________

Myth 5: Astronauts are weightless because there is no air in space.

______________________________

______________________________

______________________________

# VOCABULARY GAMES and CARDS

## Contents

# Vocabulary Games

**You can use the vocabulary cards on pages CS133–CS252 to play these games. The cards are provided for each chapter in your science textbook. Each card has a word on one side and the word's definition on the back. For some of these games, you may need to keep the definition hidden from view. Or, you may need to use a photocopy of one side of the page.**

## Make a Board Game

**You will need**

vocabulary cards, number cube, markers, poster board or file folder, playing pieces

**Grouping** small group or partners

1. Make a simple path on a file folder. Write directions randomly on the spaces, such as *Take another turn* or *Leap ahead 2.* Place the word cards face up in a pile on the board so that the words are showing but the definition is hidden.
2. The first player picks a word card from the pile, reads the word, and uses it in a sentence. The definition is read to make sure the word was used correctly. If the player was correct, he or she tosses the number cube, moving a playing piece ahead that number of spaces. He or she follows any directions on the space.
3. The player who reaches the end of the game path first is the winner!

## Concentration

**You will need**

one-sided word cards, matching one-sided definition cards

**Grouping** small group or pairs

1. On a flat surface, place the shuffled word and definition cards face down.
2. Players take turns trying to match a word to its definition by turning over two cards.
3. If the two cards match, that player keeps the cards and takes another turn. If the cards do not match, the player turns the cards face down and the other player takes a turn. The player who finds the most matches is the winner.

## Metamorphosis

**You will need**

drawing materials, vocabulary cards, paper and pencil

**Grouping** small groups or pairs

1. A metamorphosis is a change in the shape or characteristics of an organism's body. In this activity, players will be changing a vocabulary word, one letter at a time, to form new words.
2. Choose one vocabulary word to transform. For example, if the vocabulary word is wave, the following new words could be formed in a chain by changing one letter: *wave, wane, want, went, lent, lens.*
3. Players then swap chains, listing the first and last words of the chain only, with dashes used for the other words in the chain. You may wish to draw pictures as clues if needed for the missing words in the chain.

## Analogies

**You will need**

vocabulary cards, paper and pencil, chalk

**Grouping** large or small groups

1. In an analogy, two pairs of words are related to each other in the same way. To complete an analogy, the words in the second pair must relate to each other in the same way as the words in the first pair. For example, *hot is to cold as tall is to short.* Both pairs of words are opposites. *Collie is to dog as rose is to flower.* Both pairs of words give examples. *Toe is to foot as branch is to tree.* Both pairs of words relate a part to a whole.
2. With a partner, write an analogy using one or more vocabulary words from the words cards. Rewrite the analogy on the chalkboard, leaving out the final element of the analogy. Be sure the missing final element is a vocabulary word.
3. After each team writes one analogy on the chalkboard, points will be given to the first team that solves the analogy correctly. The group that has the most points is the winning team.

## Jumbles

**You will need**

vocabulary cards, chalk, paper and pencil

**Grouping** small group or pairs

1. To jumble a word is to scramble the letters of that word. For example, *axagly* is the jumbled form of the word *galaxy*.
2. Create jumbles for each of the vocabulary words found on the word cards. For each jumbled word, you may wish to provide its definition as a clue for that word. Provide other players dashes for each of the letters found in a word.
3. Exchange jumbles with another player. The person to solve all of the jumbles first is the winner.

## What's the Question?

**You will need**

vocabulary cards, paper and pencil

**Grouping** large or small groups

1. One player will act as the moderator, or "game show host". He or she will use both sides of the vocabulary cards while the other players use just the side with the words.
2. The other players should spread their cards out in front of them, word side up, so they can be easily read. The moderator will give one clue by reading a word's definition to the group. For example, "It's a force that pulls us toward Earth, so we don't fly off into space."
3. The players will tap their pencils on a desk if they know the answer. The first player to tap is called on by the moderator. That player must quickly answer in the form of a question, for example, "What is gravity?"
4. Players receive points for each correct answer. The player with the most points wins the game.

1
absorb
2
acid

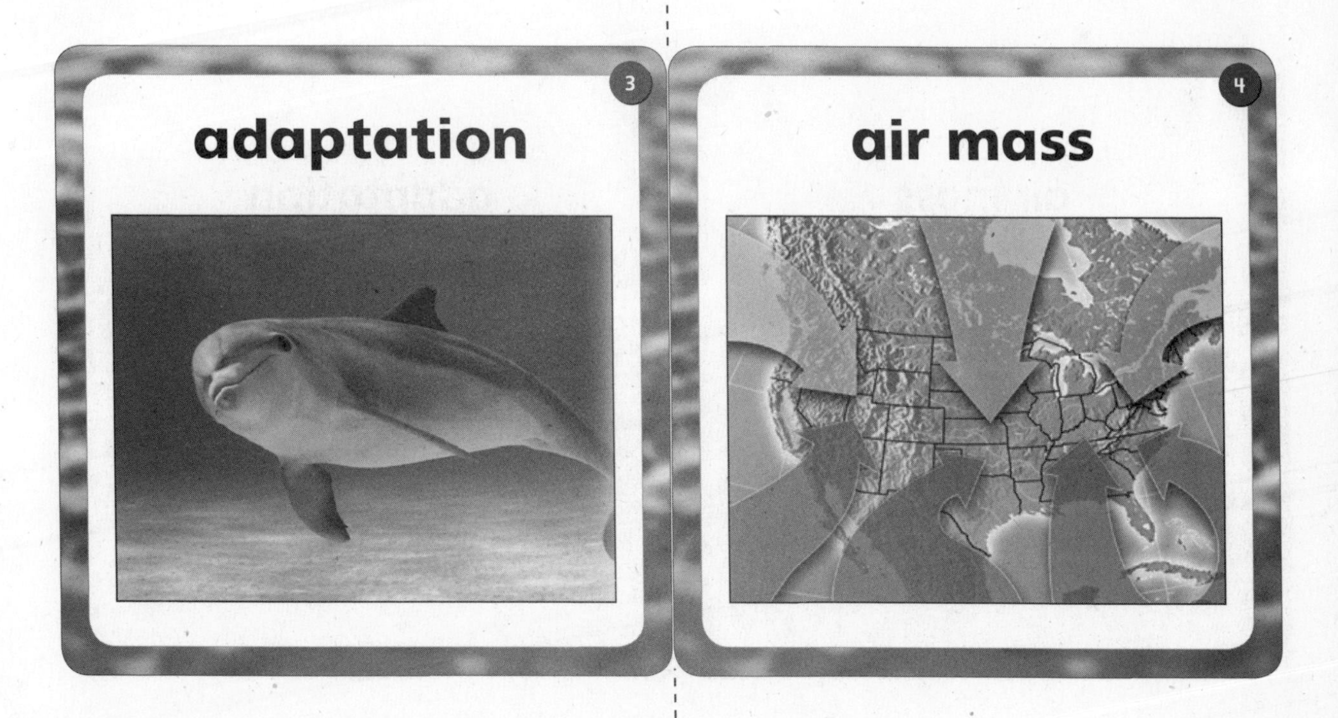
3
adaptation
4
air mass

## absorb

[ab•ZAWRB]

To take in.

Sponges *absorb* liquid easily.

## acid

[AS•id]

A chemical compound that turns blue litmus paper red and has a pH of less than 7.

The juice from an orange is a mild *acid*.

## adaptation

[ad•uhp•TAY•shuhn]

A trait or characteristic that helps an organism survive.

The dolphin's flippers are an *adaptation* that helps it swim.

## air mass

[AIR MAS]

A large body of air that has similar temperature and humidity throughout.

The blue arrows represent cool *air masses*.

5

**air pressure**

6

**alloy**

7

**anemometer**

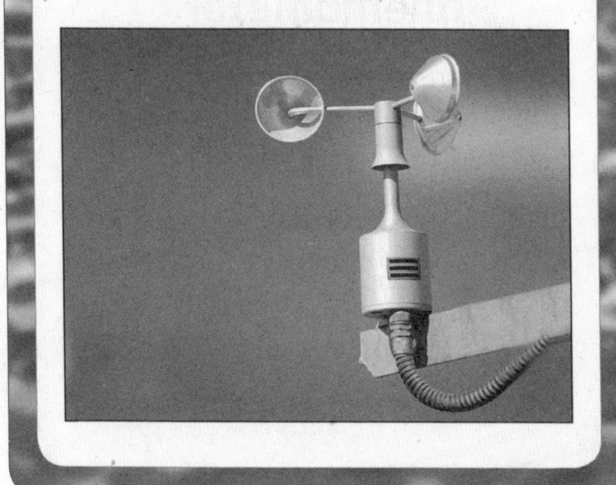

8

**aqueduct**

6

## alloy

[AL•OY]

A solid solution in which a metal or a nonmetal dissolves in a metal.

This statue is made of an *alloy* called bronze.

5

## air pressure

[air PRESH•er]

The weight of the atmosphere pressing down on Earth.

*Air pressure* changes with altitude.

8

## aqueduct

[AK•wuh•duhkt]

A pipe or channel that is used to transport water.

You can see *aqueducts* built by ancient Romans throughout western Europe.

7

## anemometer

[an•uh•MAHM•uht•er]

An instrument for measuring wind speed.

Wind makes an *anemometer* spin.

9

**array**

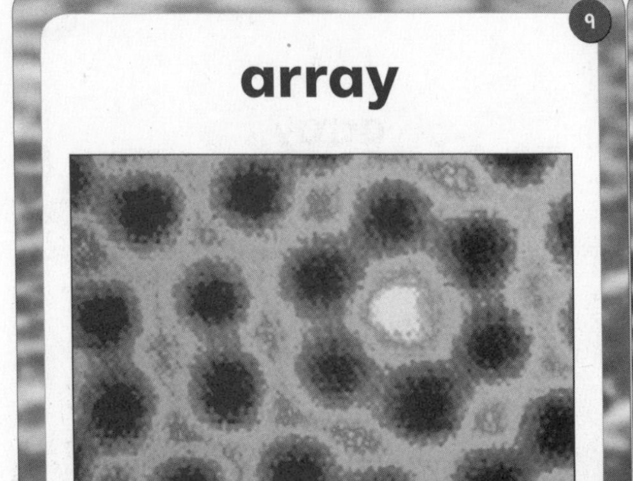

10

**artery**

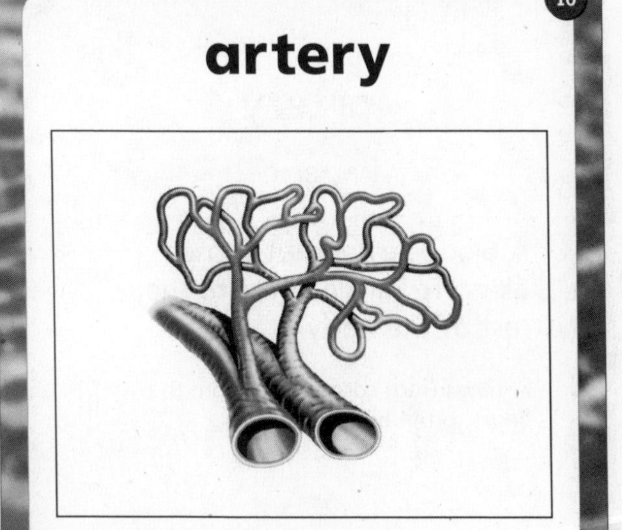

11

**asteroid**

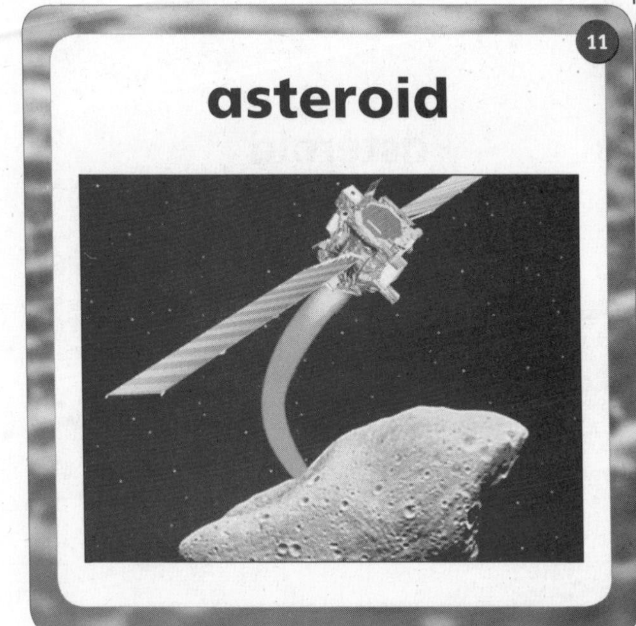

12

**atmosphere**

10

## artery

[ART•er•ee]

A blood vessel that carries blood from the heart to the rest of the body.

*Arteries* (red) carry blood from the heart to the body.

9

## array

[uh•RAY]

A pattern of atoms.

This *array* of atoms can be seen only through a very powerful microscope.

12

## atmosphere

[AT•muhs•fir]

The blanket of air surrounding Earth.

Earth's *atmosphere* has several layers.

11

## asteroid

[AS•ter•oyd]

A piece of rock and metal that orbits the sun.

Some *asteroids* are pieces of rock from collisions of larger objects in space.

13
atom

14
atomic number
6
C
Carbon

15
balance

16
barometer

14

## atomic number

[uh•TAHM•ik NUHM•ber]

The number of protons in an atom.

The *atomic number* of carbon is 6.

13

## atom

[AT•uhm]

The smallest unit of an element, that has the properties of that element.

Nearly all *atoms* have neutrons.

16

## barometer

[buh•RAHM•uht•er]

An instrument for measuring air pressure.

The original *barometer* used mercury in a glass tube to measure air pressure.

15

## balance

[BAL•uhns]

A tool that measures an object's mass.

The *balance* shows that the masses of these objects are equal.

17

**barometric pressure**

18

**base**

19

**binoculars**

20

**bladder**

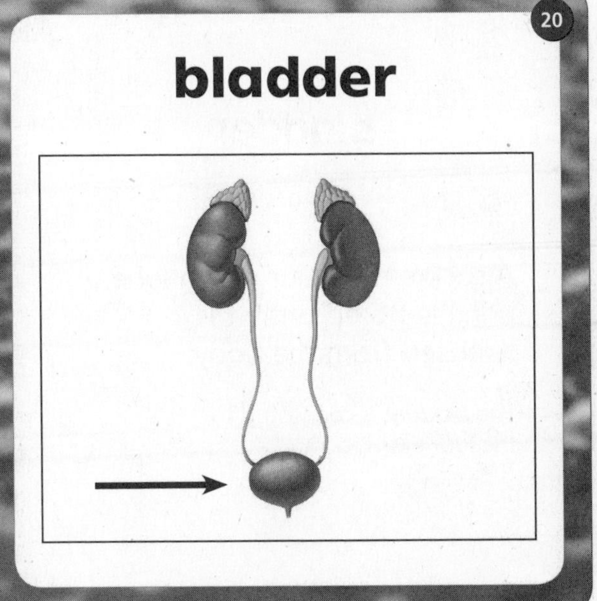

18

## base

[BAYS]

A chemical compound that turns red litmus paper blue and has a pH of more than 7.

Soap is a mild *base*.

17

## barometric pressure

[bair•uh•MEH•trik PRESH•er]

The weight of the atmosphere pressing down on Earth; also called air pressure.

When the *barometric pressure* drops quickly, you can expect a storm.

20

## bladder

[BLAD•er]

A saclike muscular organ where urine is stored until it is released from the body.

The *bladder* is connected to the kidneys.

19

## binoculars

[by•NAHK•yuh•lerz]

A device for looking at distant object that magnifies what is seen using a lens for each eye.

You can use *binoculars* to see things that are far away.

21

## blizzard

22

## blood circulation

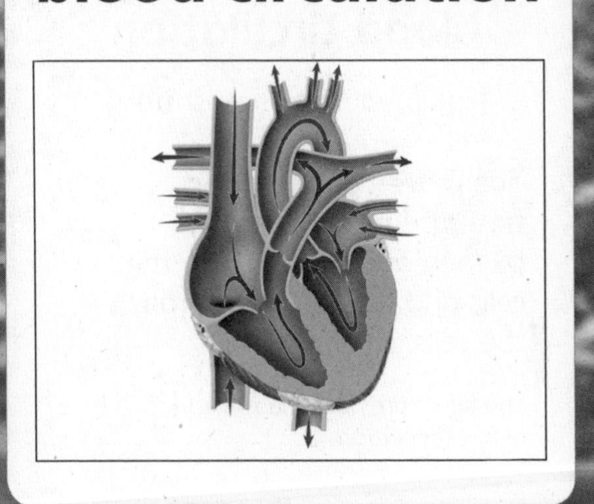

23

## boil

24

## brittle

22

## blood circulation

[BLUHD ser•kyoo•LAY•shuhn]

The movement of blood through the body, taking oxygen and nutrients to the cells and wastes away from the cells.

The heart provides the force for *blood circulation*.

21

## blizzard

[BLIZ•erd]

A severe snowstorm.

It is hard to see in a *blizzard*.

24

## brittle

[BRIT•uhl]

Able to be broken or crushed easily.

Chalk is *brittle*.

23

## boil

[BOYL]

When a substance changes from a liquid to a gas.

Water *boils* at 100°C (212°F).

25

**capillary**

26

**carbon dioxide**

27

**cardiovascular**

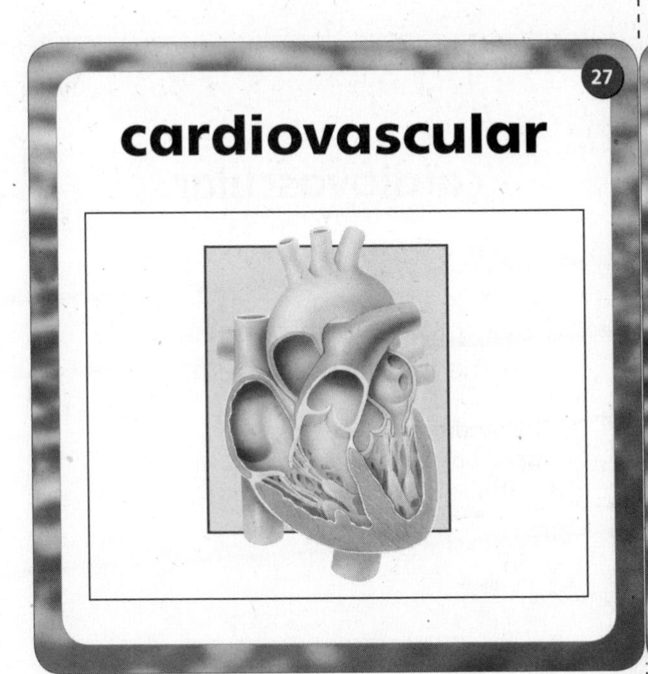

28

**cell**

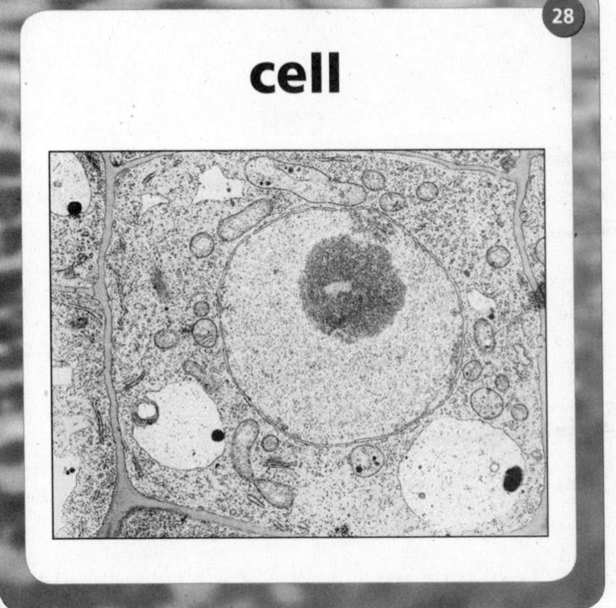

## carbon dioxide

[KAR•buhn dy•AHK•syd]

A molecule formed from one atom of carbon and two atoms of oxygen.

*Carbon dioxide* is the compound in dry ice.

26

## capillary

[KAP•uh•lair•ee]

One of tiny blood vessels that exchange materials between the blood and body cells.

You have many *capillaries* in your skin.

25

## cell

[sel]

The basic unit of structure and function of all living things.

Plants *cells* have a cell wall.

28

## cardiovascular

[kar•dee•oh•VAS•kyoo•ler]

Having to do with the heart and the circulatory system.

The condition of your heart is important to your *cardiovascular* health.

27

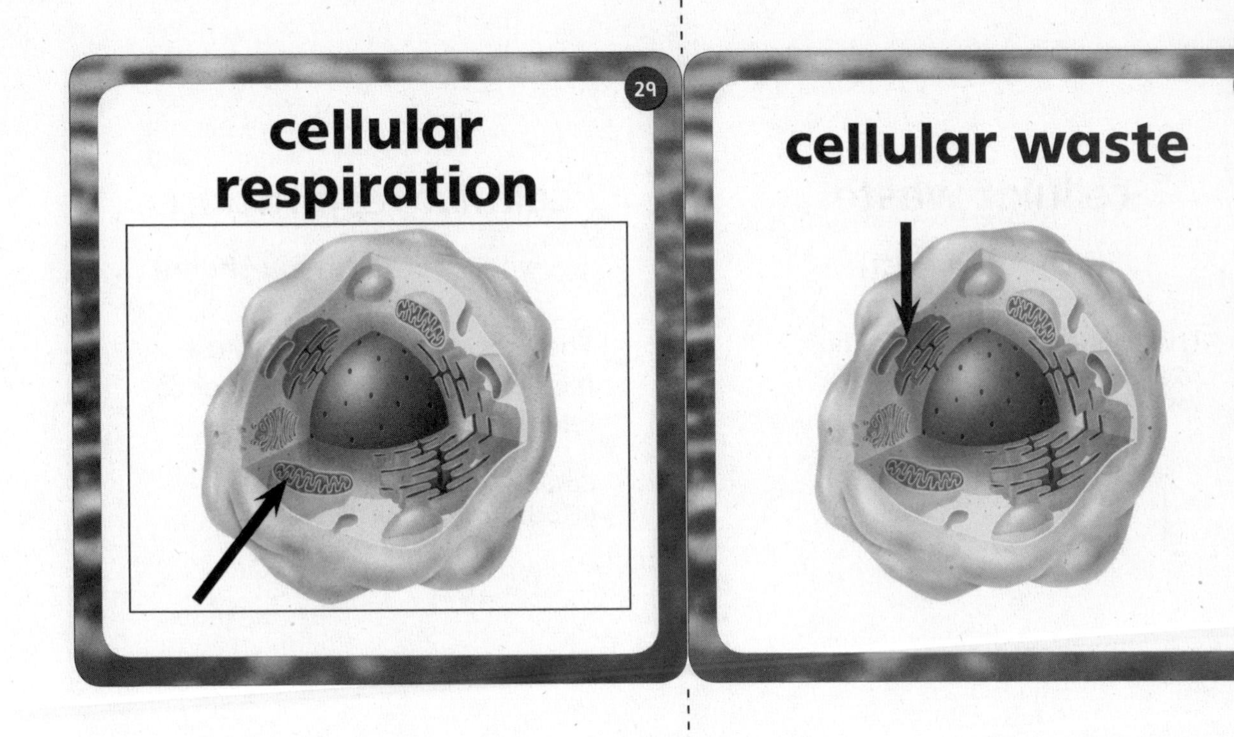
29
cellular respiration
30
cellular waste

31
chaos theory
32
chemical change

30

## cellular waste

[SEL•yoo•ler WAYST]

The product of cell functions.

Vesicles store *cellular waste*.

29

## cellular respiration

[SEL•yoo•ler res•puh•RAY•shuhn]

The process by which cells use oxygen to break down sugar to release energy.

*Cellular respiration* takes place in the mitochondria.

32

## chemical change

[KEM•ih•kuhl CHAYNJ]

A change in which a substance or two becomes a new substance or two.

Burning is one kind of *chemical change*.

31

## chaos theory

[KAY•ahs THEE•uh•ree]

The idea that very small changes can have major effects on a system.

An illustration of *chaos theory* is the effect that one falling domino has on the others in a row.

33

**chemical compound**

34

**chemical property**

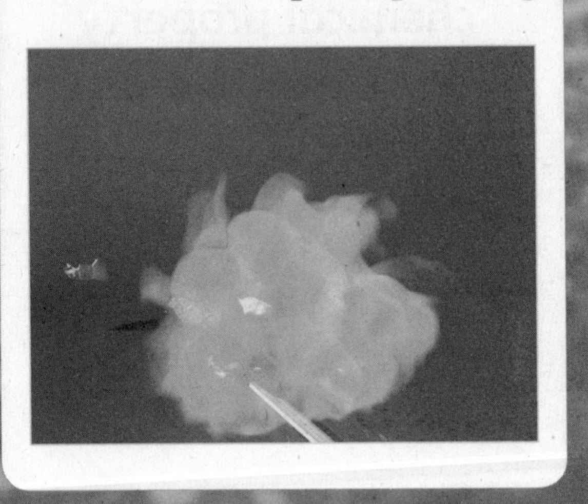

35

**chemical reaction**

36

**chlorophyll**

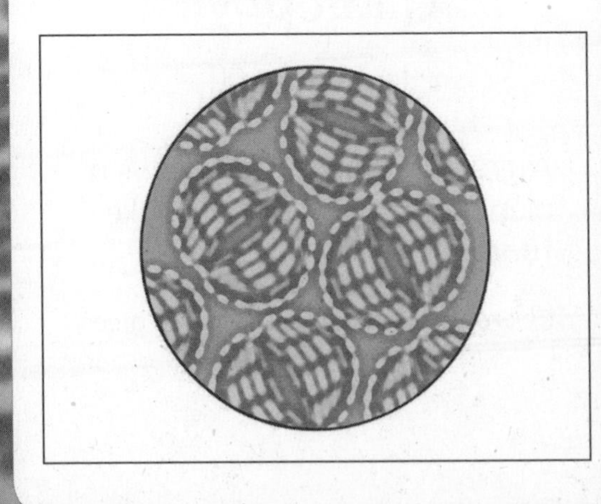

34

## chemical property

[KEM•ih•kuhl PRAHP•er•tee]

A property that involves the ability of a substance to react with other materials and form new substances.

Flammability is one kind of *chemical property*.

33

## chemical compound

[KEM•ih•kuhl KAHM•pownd]

A substance made of two or more different elements.

Water is a *chemical compound* because it is made of hydrogen and oxygen.

36

## chlorophyll

[KLAWR•uh•fil]

A green pigment that allows a plant cell to use light to make food.

*Chlorophyll* is what makes leaves green.

35

## chemical reaction

[KEM•ih•kuhl ree•AK•shuhn]

A change in which one or more new substances are formed.

Rusting is one kind of *chemical reaction*.

37

## circulation

38

## circulatory system

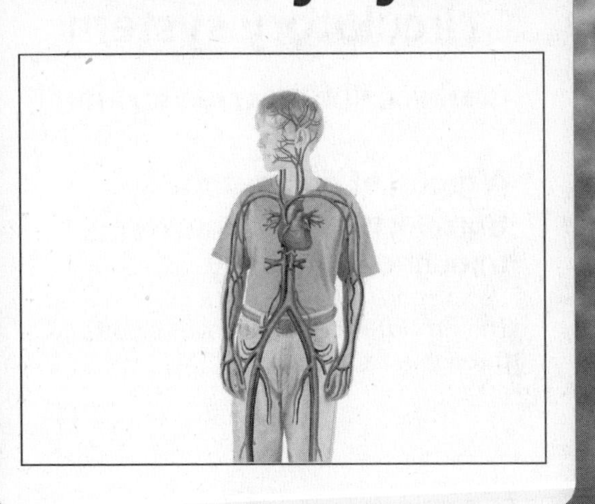

39

## classify

40

## cloud

38

## circulatory system

[SER•kyoo•luh•tawr•ee SIS•tuhm]

A group of organs that transports needed materials throughout the body.

The *circulatory system* moves blood throughout the body.

37

## circulation

[ser•kyoo•LAY•shuhn]

The movement of something from place to place (water and air around Earth).

The water cycle is the *circulation* of water from the atmosphere to Earth's surface and back to the atmosphere.

40

## cloud

[KLOWD]

Water that has either condensed on dust particles in the air, or frozen at a high altitude.

The types of *clouds* are related to different kinds of weather.

39

## classify

[KLAS•uh•fy]

To group or organize objects or events into categories based on similar criteria.

This student is *classifying* objects.

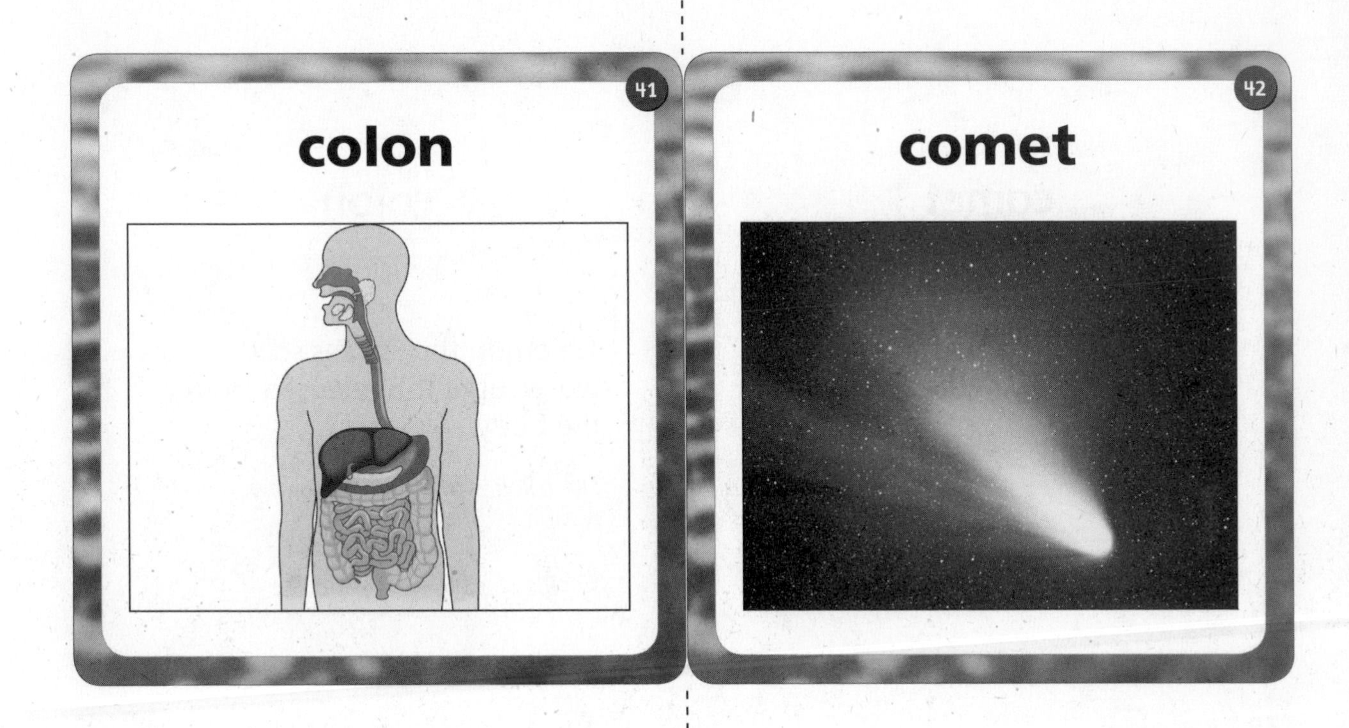
41
colon
42
comet

43
compound
Baking Soda
Vinegar

44
conclusion

42

## comet

[KAHM•it]

A ball of ice, rock, and frozen gases that orbits the sun.

A *comet's* orbit around the sun is usually irregular.

41

## colon

[KOH•luhn]

An organ that stores solid waste until it is released from the body.

The *colon* is part of the digestive system.

44

## conclusion

[kuhn•KLOO•zhuhn]

A decision you make based on information.

These students are using information from an experiment to draw a *conclusion*.

43

## compound

[KAHM•pownd]

A substance made of two or more different elements.

Baking soda and vinegar are *compounds*.

45

## condensation

46

## conductivity

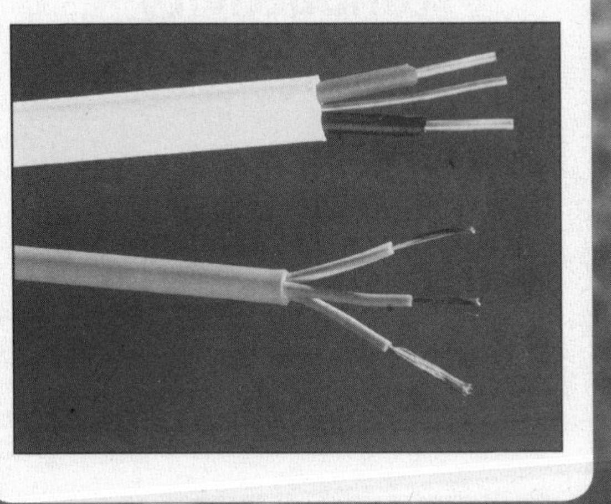

47

## conservation

48

## control variable

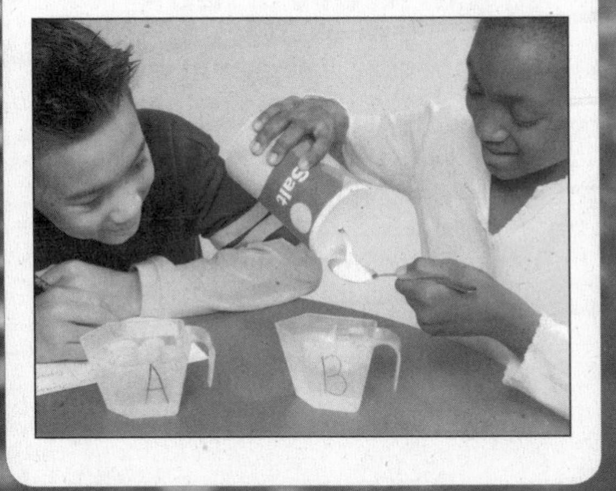

46

## conductivity

[kahn•duhk•TIV•uh•tee]

The ability of a metal to transfer energy easily.

These wires are made of metal that has very high *conductivity*.

45

## condensation

[kahn•duhn•SAY•shuhn]

The process by which a gas changes into a liquid.

Rain results from *condensation*.

48

## control variable

[kuhn•TROHL VAIR•ee•uh•buhl]

The parts of an investigation that you can control.

It is important to keep track of your *control variables*.

47

## conservation

[kahn•ser•VAY•shuhn]

The preserving and protecting of a resource.

Protecting resources is a part of *conservation*.

49

**convection**

50

**convection currents**

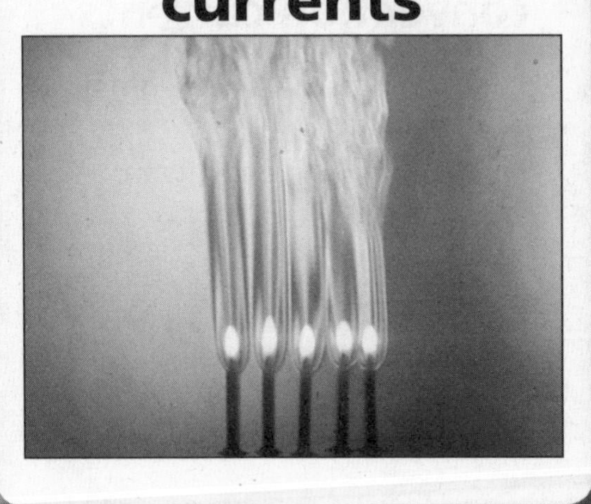

51

**creek**

52

**crescent moon**

50

## convection currents

[kuhn•VEK•shuhn KER•uhnts]

The upward and downward movements of a gas or a liquid.

These candles heat the air and cause it to move in a *convection current*.

49

## convection

[kuhn•VEK•shuhn]

Circular movement in a liquid or gas, resulting from regions of different temperatures and different densities.

*Convection* in the atmosphere produces rain clouds.

52

## crescent moon

[KRES•uhnt moon]

The moon phase just after or just before a new moon.

There are two *crescent moons* in a month.

51

## creek

[KREEK]

A narrow and shallow river.

Water can move quickly, even in a small *creek*.

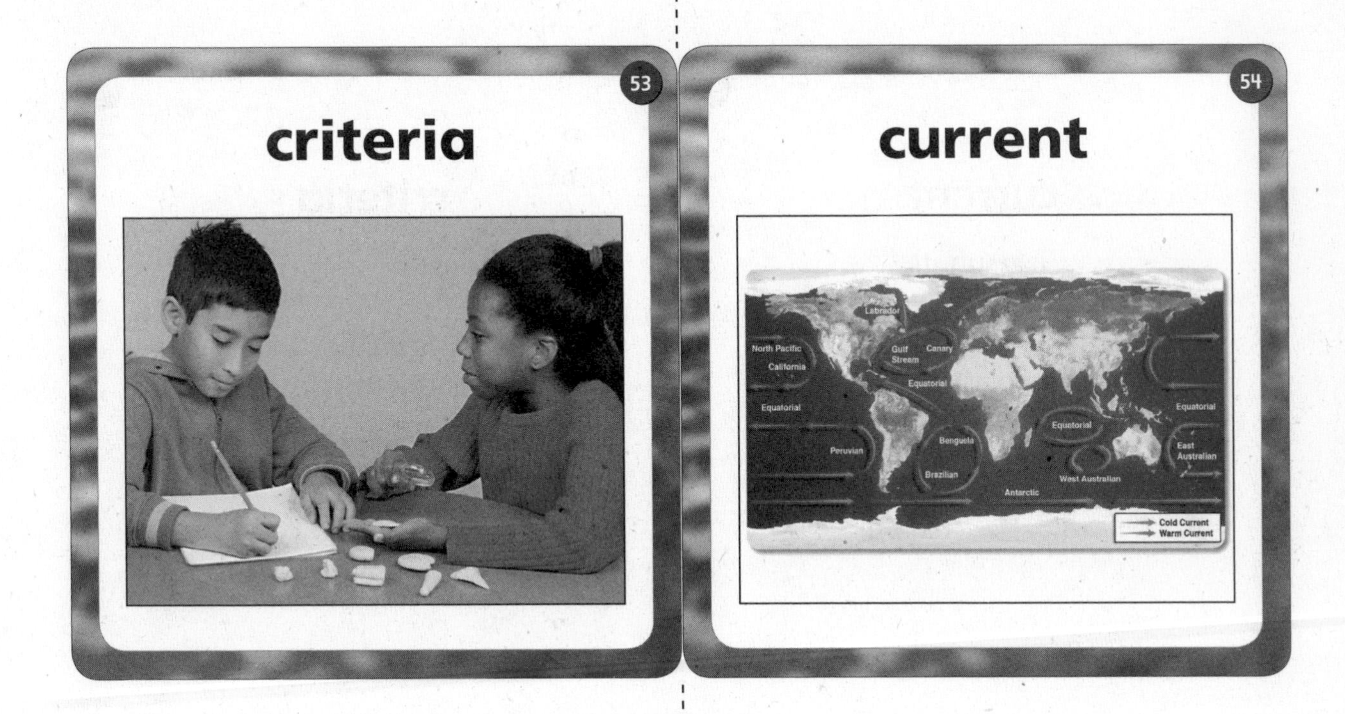
53
criteria
54
current
Labrador
North Pacific
California
Gulf Stream
Canary
Equatorial
Equatorial
Equatorial
Equatorial
Peruvian
Benguela
Brazilian
Antarctic
West Australian
East Australian
Cold Current
Warm Current

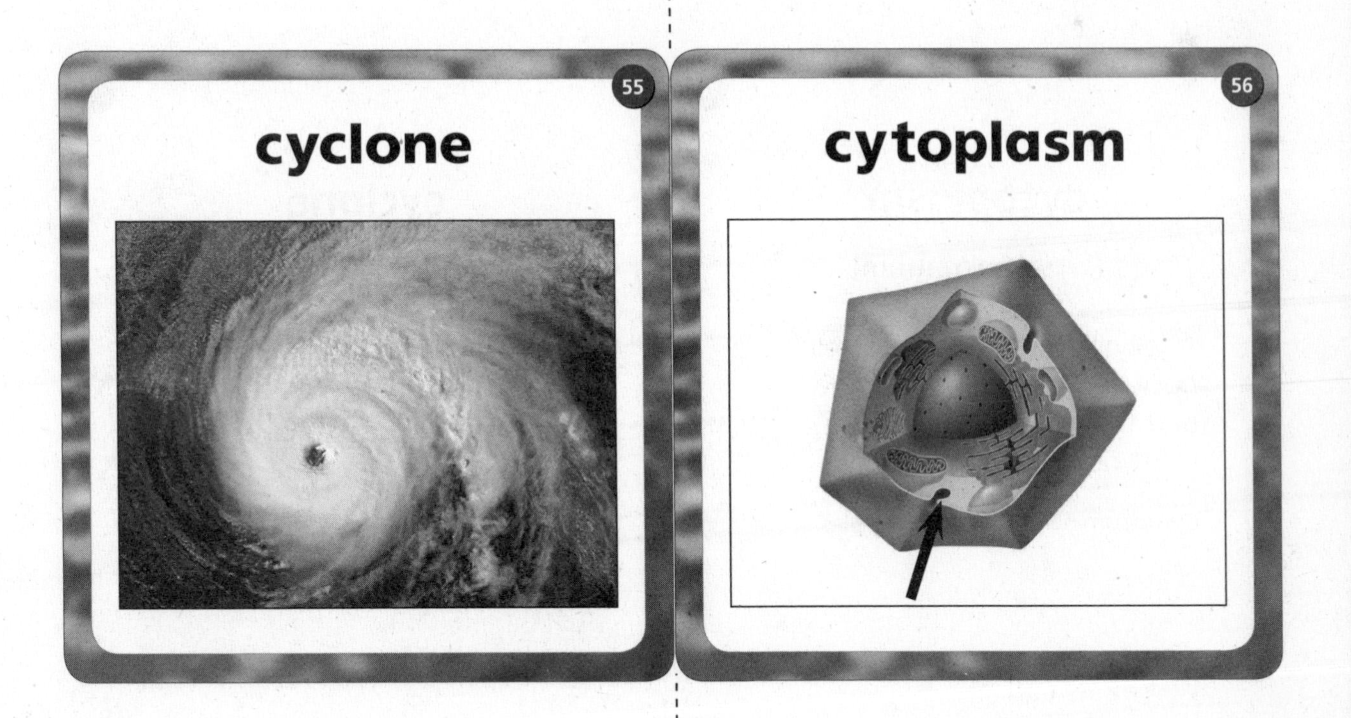
55
cyclone
56
cytoplasm

54

## current

[KER•uhnt]

A stream of water that flows like a river through the ocean.

Ocean *currents* flow in only one direction.

53

## criteria

[kry•TIR•ee•uh]

The specific qualities that allow you to group items.

The students are classifying the objects according to the *criteria* they discussed.

56

## cytoplasm

[SYT•oh•plaz•uhm]

A jellylike substance in a cell between the cell membrane and the nucleus, containing most organelles.

*Cytoplasm* helps protect organelles.

55

## cyclone

[SY•klohn]

A rapidly turning air mass.

The air in a *cyclone* turns counterclockwise in the Northern Hemisphere.

57
dam
58
deflect

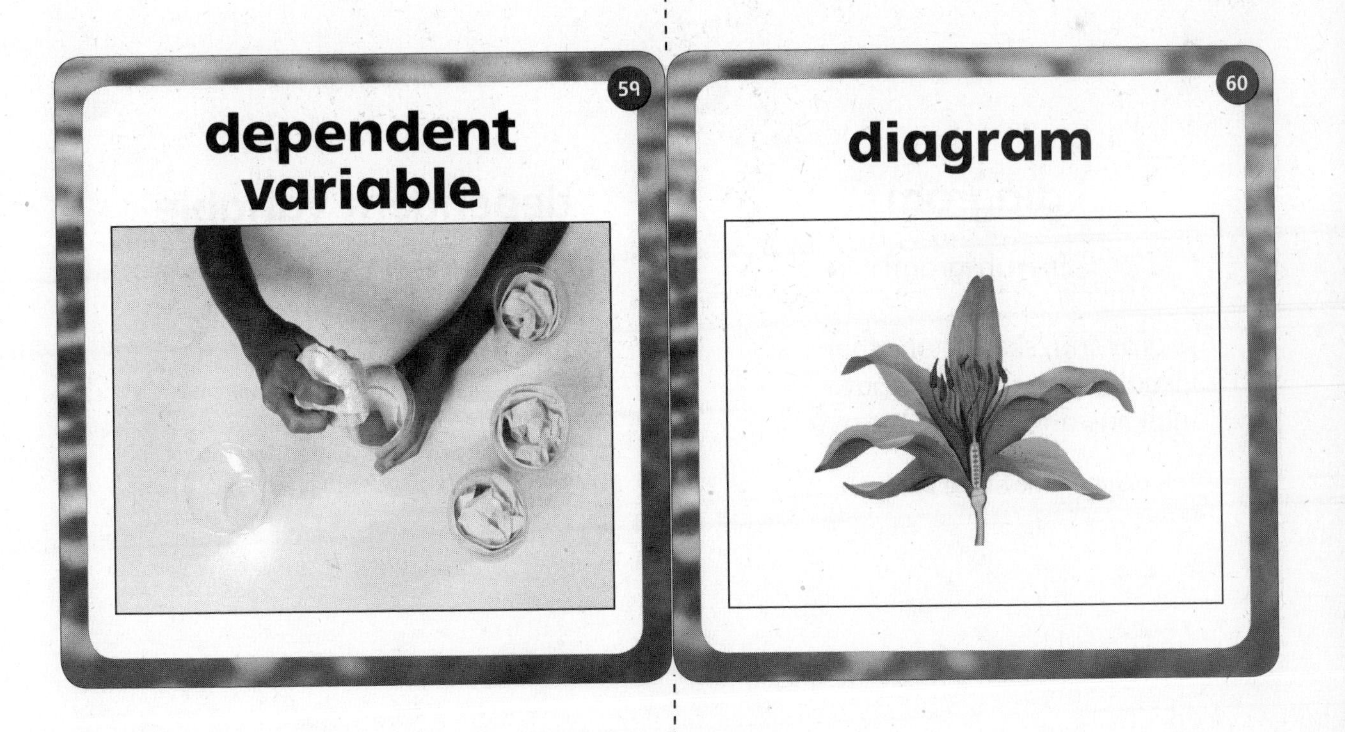
59
dependent variable
60
diagram

58

## deflect

[dee•FLEKT]

To turn something from its path, as when winds are deflected by a moutain range.

The windshield of a car *deflects* the air.

57

## dam

[DAM]

A barrier across a river, controlling its flow.

*Dams* can be natural, animal-made, or human-made.

60

## diagram

[DY•uh•gram]

A drawing, sketch, or other visual representation that explains an idea or object.

This *diagram* shows the parts of a flower.

59

## dependent variable

[dee•PEN•duhnt VAIR•ee•uh•buhl]

The part of an investigation that is out of your control.

The results of your investigation are shown by a *dependent variable*.

61

## diffusion

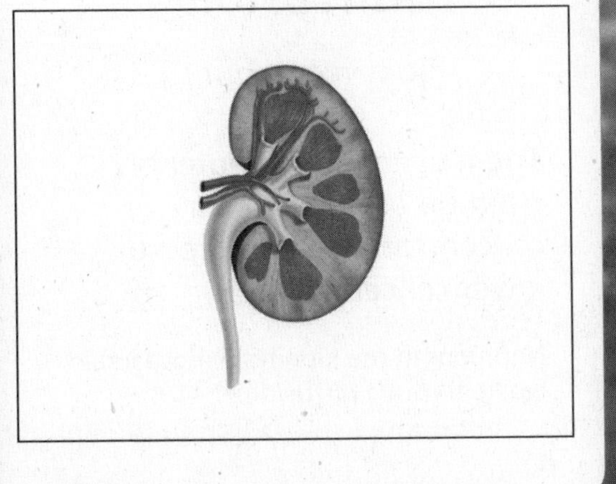

62

## digestion

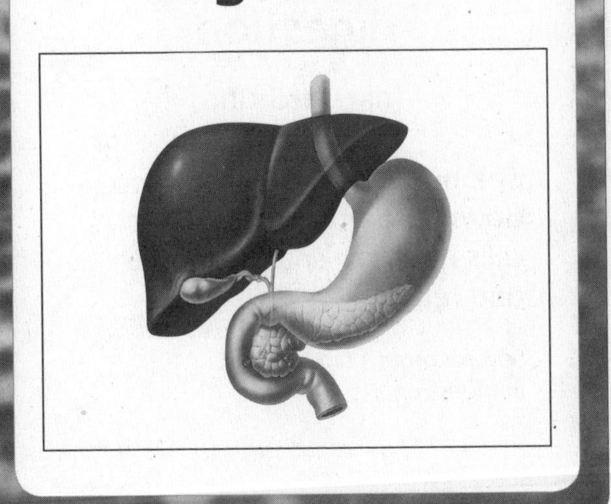

63

## digestive system

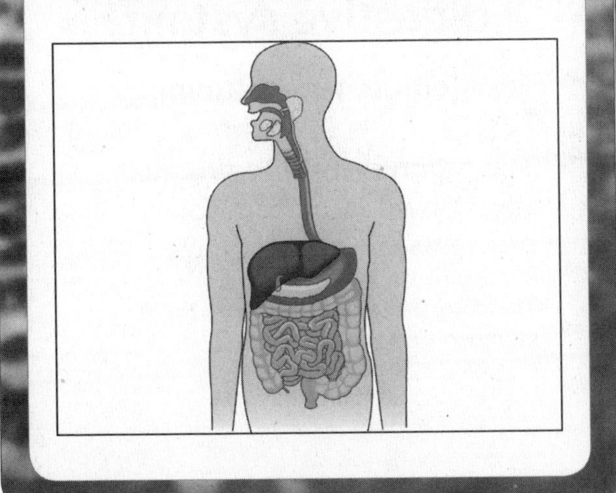

64

## ductile

62

## digestion

[dih•JES•chuhn]

The process of breaking food down into nutrients the body's cells need for energy, growth, and repair.

Several organs are involved in *digestion*.

61

## diffusion

[dih•FYOO•zhuhn]

The movement of materials from an area of higher concentration to an area of lower concentration.

Nephrons in the kidneys remove liquid waste through *diffusion*.

64

## ductile

[DUHK•tuhl]

Able to be pulled into thin strands.

*Ductile* metal is used to make wire.

63

## digestive system

[dih•JES•tive SIS•tuhm]

The organ system that breaks food down into chemical nutrients the body can use.

The *digestive system* includes the stomach and the intestines.

65
Earth

66
El Niño

67
electrical
conductivity

68
electron

66

## El Niño

[el NEEN•yoh]

Warming of South American equatorial ocean waters that leads to changing weather patterns.

On this satellite map, you can see the warm water that causes *el Niño*.

65

## Earth

[ERTH]

The planet we live on.

*Earth* looks mostly blue from space.

68

## electron

[ee•LEK•trahn]

One of the particles in an atom.

An *electron* has a negative charge.

67

## electrical conductivity

[ee•LEK•trih•kuhl kahn•duhk•TIV•uh•tee]

The ability of a metal to transfer electrons.

Silver has *electrical conductivity*.

69

# electron microscope

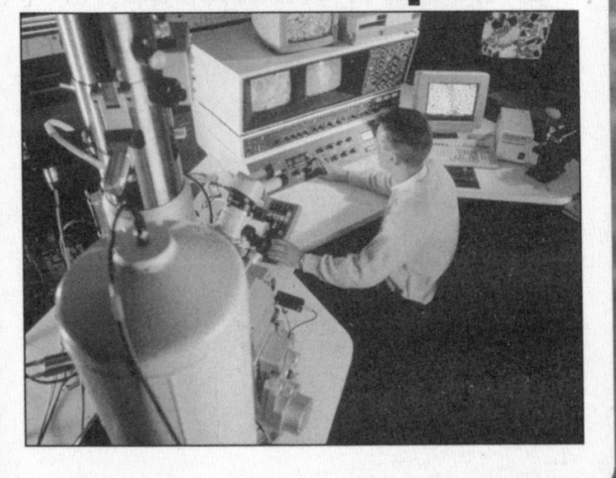

70

# element

71

# elliptical

72

# Equator

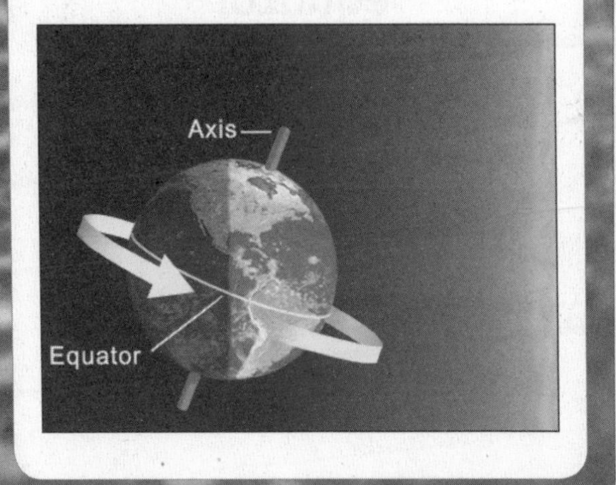

## element

[EL•uh•muhnt]

A substance made up of only one kind of atom.

Gold is an *element* because it is made of only gold atoms.

## electron microscope

[ee•LEK•trahn MY•kruh•skohp]

A microscope that uses a stream of electrons to produce images of objects.

The *electron microscope* is very powerful.

## equator

[ee•KWAYT•er]

An imaginary line around Earth equally distant from the North and South Poles.

The *equator* divides Earth into the Northern and Southern Hemispheres.

## elliptical

[eh•LIP•tuh•kuhl]

Oval shaped.

The orbits of most planets are *elliptical*.

73

## erosion

74

## esophagus

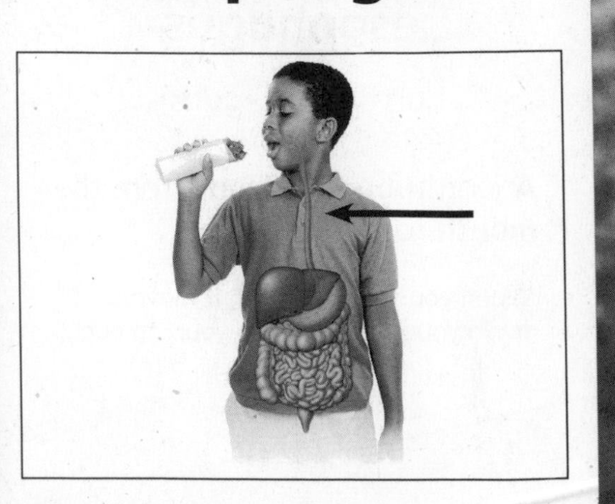

75

## evaporate

76

## evaporation

## esophagus

[ih•SAHF•uh•guhs]

A long tube that leads from the mouth to the stomach.

When you swallow food, it moves down your *esophagus* to your stomach.

74

## erosion

[ee•ROH•zhuhn]

The process of moving sediment by wind, moving water, or ice.

This gully was formed by *erosion*.

73

## evaporation

[ee•vap•uh•RAY•shuhn]

The process by which liquid changes into a gas.

The water level of this lake has dropped because of *evaporation*.

76

## evaporate

[ee•VAP•uh•rayt]

To change from a liquid to a gas.

The sun's heat causes ocean water to *evaporate*.

75

77

## evidence

78

## excretion

79

## excretory system

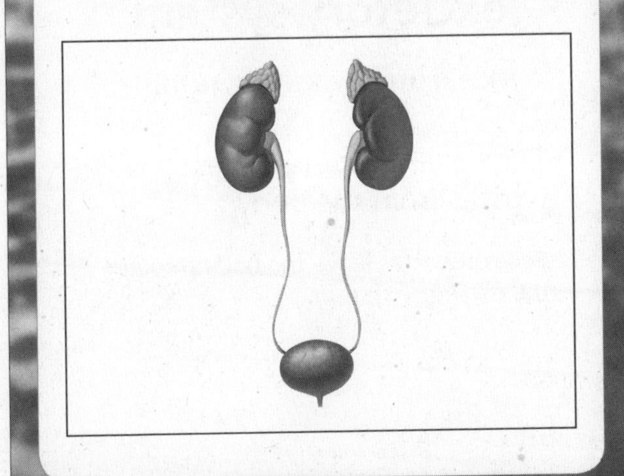

80

## experiment

## 78

### excretion

[eks•KREE•shuhn]

The removal of wastes from the body.

Sweating is one kind of *excretion*.

## 77

### evidence

[EV•uh•duhns]

Information, collected during an investigation, to support a hypothesis.

A scientist gathers *evidence* from an experiment.

## 80

### experiment

[ek•SPAIR•uh•muhnt]

A procedure you carry out under controlled conditions to test a hypothesis.

*Experiments* are an important part of the scientific method.

## 79

### excretory system

[EKS•kruh•tawr•ee SIS•tuhm]

The system that removes wastes from the body.

Cellular wastes leave the body through the *excretory system*.

81

**ferment**

82

**fermentation**

83

**flood basin**

84

**fog**

## fermentation

[fer•muhn•TAY•shuhn]

The process that releases energy from sugar in the absence of oxygen.

*Fermentation* bubbles are what cause bread dough to rise.

82

## ferment

[FUR•ment]

To release energy from sugar in the absence of oxygen.

*Yeast ferments making bread dough rise.*

81

## fog

[FAWG]

A cloud that forms near the ground.

*Fog* can make it hard for drivers to see the road ahead.

84

## flood basin

[FLUHD BAY•suhn]

An area of land that "catches" the runoff from urban areas in a human-made lake.

When people build near a *flood basin*, their houses are at risk of being flooded.

83

85

## force

86

## forecast

87

## formula

88

## freeze

86

## forecast

[FAWR•kast]

The prediction of future weather.

Meteorologists make weather *forecasts*.

85

## force

[FAWRS]

A push or pull that causes an object to move, stop, or change direction.

*Forces* affect the movement of objects.

88

## freeze

[FREEZ]

To change from a liquid to a solid.

Water *freezes* at 0°C (32°F).

87

## formula

[FAWRM•yuh•luh]

Symbols that show how many atoms of each element are present.

The *formula* for water is $H_2O$.

89

**freezing**

90

**front**

91

**full moon**

92

**fusion**

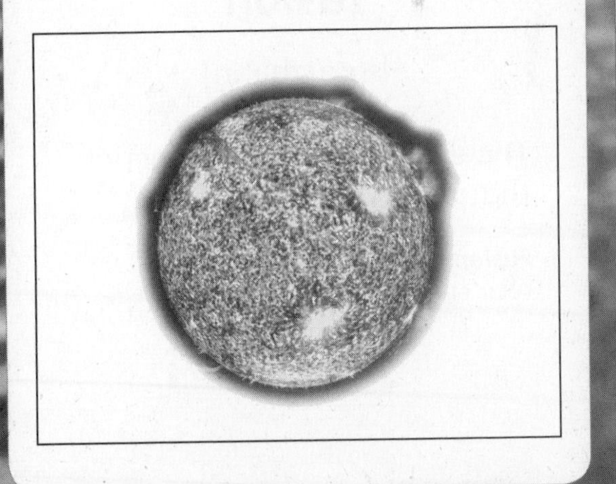

90

## front

[FRUHNT]

A place where two air masses meet.

Sometimes you can locate a *front* by watching the clouds.

89

## freezing

[FREEZ•ing]

Having an air temperature below 0°C (32°F).

Dripping water can form icicles when the temperature outdoors is *freezing*.

92

## fusion

[FYOO•zhuhn]

The energy–producing reaction that occurs inside of stars.

*Fusion* inside the sun produces solar energy.

91

## full moon

[FUL MOON]

The moon phase in which all of the moons surface facing Earth is visible.

A second *full moon* in any month is called a blue moon.

93

**gas**

94

**glacier**

95

**graduated cylinder**

96

**gravity**

94

## glacier

[GLAY•sher]

A huge sheet of moving ice.

*Glaciers* form in places where new snow falls faster than old snow melts.

93

## gas

[GAS]

The state of matter that does not have a definite shape or volume.

These balloons contain helium *gas*.

96

## gravity

[GRAV•ih•tee]

The force that pulls objects toward Earth.

On a roller coaster, you experience the effects of *gravity*.

95

## graduated cylinder

[GRA•joo•ay•tuhd SIL•uhn•der]

A tool used to make quantitative observations of the volume of liquids.

*Graduated cylinders* come in several sizes.

97

# groundwater

98

# hail

99

# heart

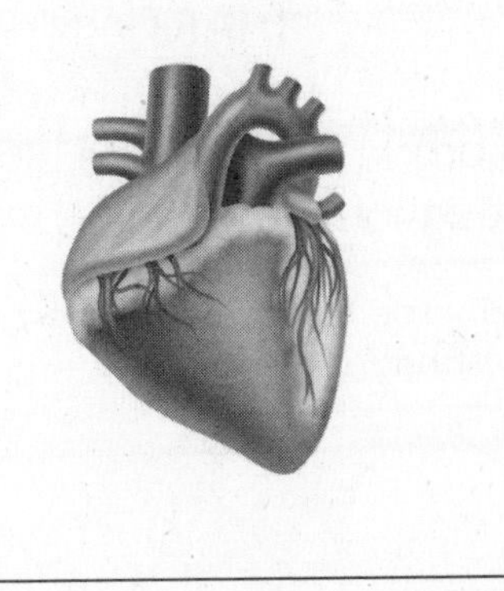

100

# helium

98

## hail

[HAYL]

Raindrops that freeze, are coated with more water, refreeze, and then fall as pieces of ice.

*Hail* sometimes falls in summer thunderstorms.

97

## groundwater

[GROWND•wawt•er]

Water that is located in soil and rocks below Earth's surface.

You can pump *groundwater* from a well.

100

## helium

[HEE•lee•uhm]

A gas that is the product of fusion inside of stars.

*Helium* is lighter than air, so it makes balloons float.

99

## heart

[HART]

An organ that pumps blood throughout the body.

The muscle tissue in your *heart* is very strong.

101

## high pressure

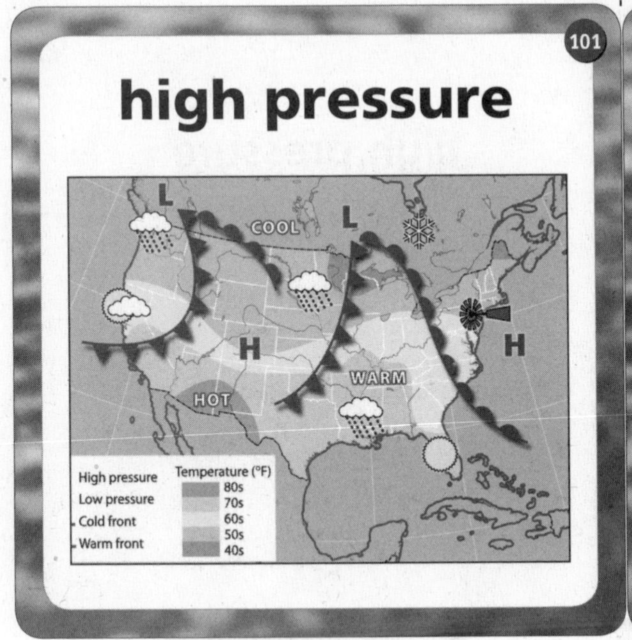


102

## humidity

103

## hurricane

104

## hydrogen

102

## humidity

[hyoo•MID•uh•tee]

A measurement of the amount of water vapor in the air.

When the *humidity* is high, your sweat evaporates slowly.

101

## high pressure

[HY•PRESH•er]

An area of dense, cold air.

Weather maps often show areas of *high pressure* with the letter "H".

104

## hydrogen

[HY•druh•juhn]

A reactant of fusion inside of stars.

Fusion can change *hydrogen* to helium.

103

## hurricane

[HER•ih•kayn]

A large, rotating tropical storm system with wind speeds of at least 119 km/hr (74 mi/hr).

*Hurricanes* form over warm ocean waters.

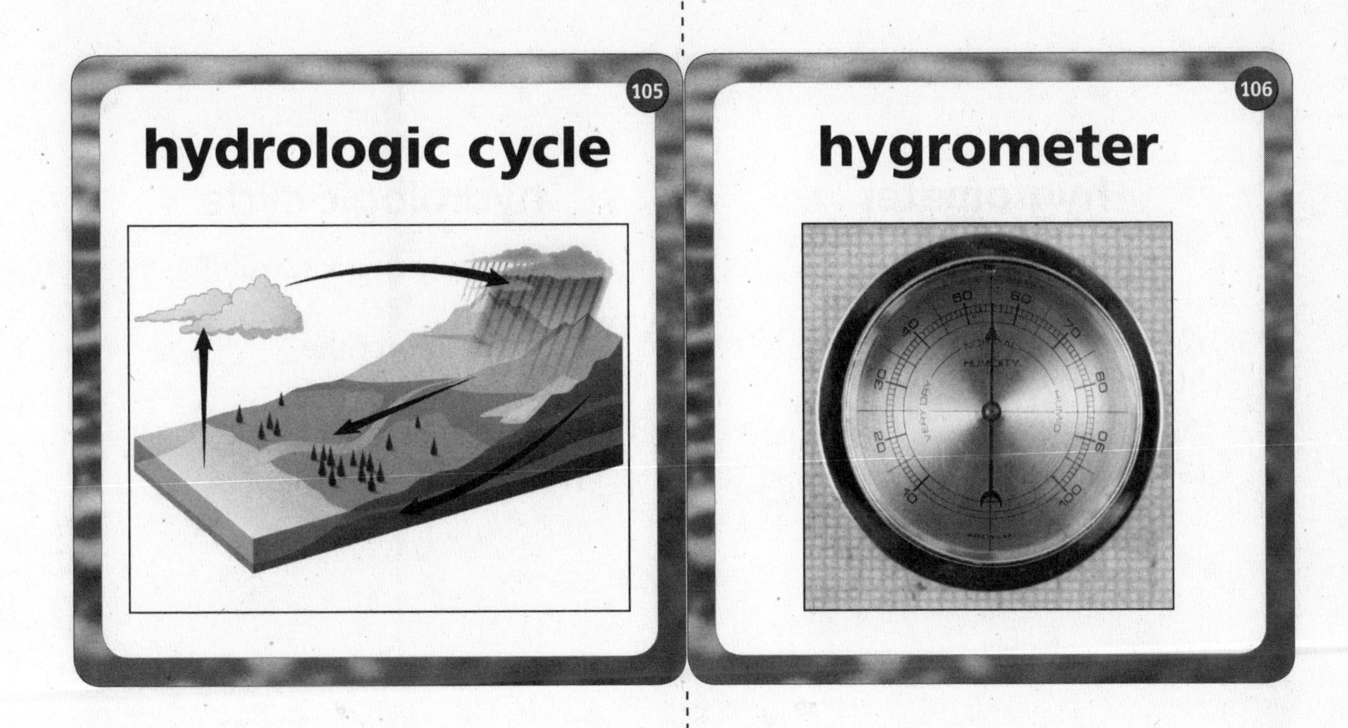
105
hydrologic cycle
106
hygrometer

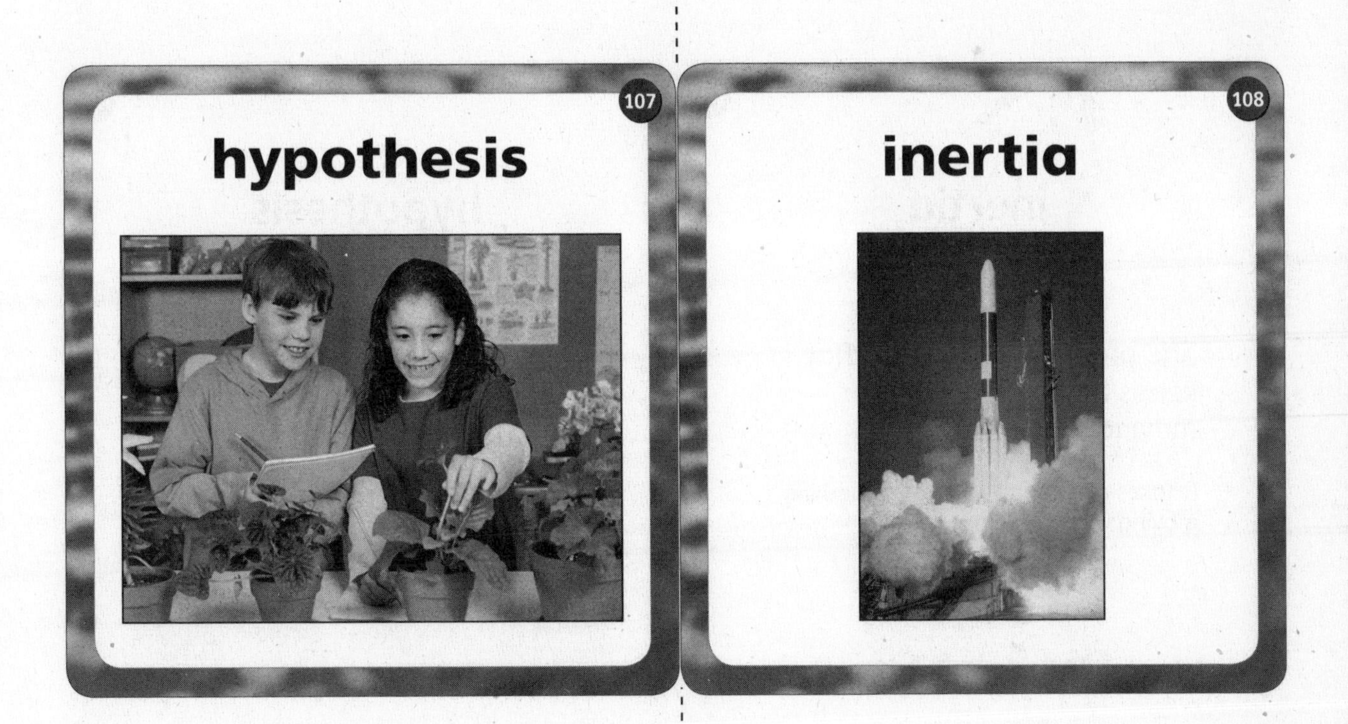
107
hypothesis
108
inertia

106

## hygrometer

[hy•GRAHM•uht•er]

An instrument for measuring humidity.

A *hygrometer* shows the relative humidity.

105

## hydrologic cycle

[hy•druh•LAHJ•ik SY•kuhl]

Another term for the water cycle.

The *hydrologic cycle* and the water cycle are the same thing.

108

## inertia

[in•ER•shuh]

The property of matter that keeps an object at rest or moving in a straight line.

It takes more force to start something moving due to *inertia*.

107

## hypothesis

[hy•PAHTH•uh•sis]

A statement that provides a testable possible answer to a scientific question.

These students are testing a *hypothesis* with their experiment.

109
inquiry
110
intestine

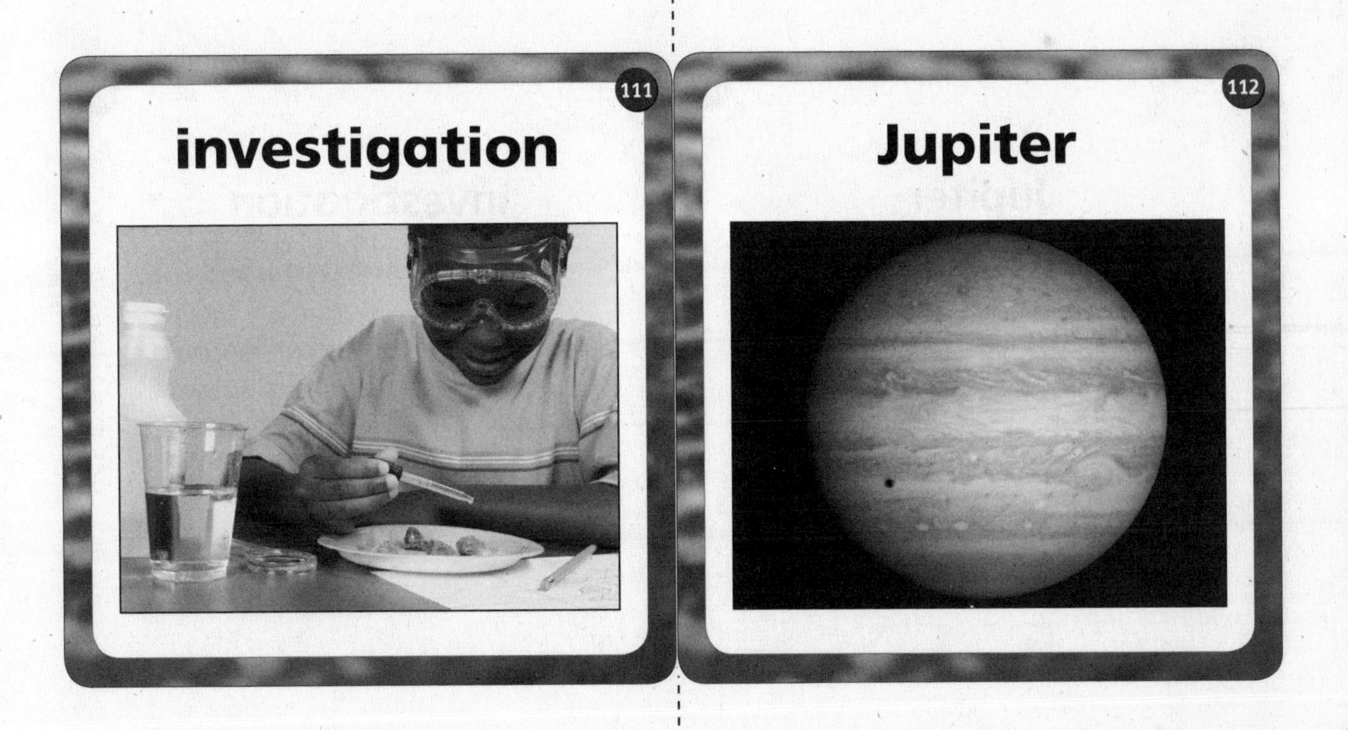
111
investigation
112
Jupiter

110

## intestine

[in•TES•tuhn]

Two connected tubes, the small and large intestines, leading from the stomach that help the body absorb water and nutrients and help rid the body of waste.

Food moves from your stomach to your small *intestine*.

109

## inquiry

[IN•kwer•ee]

An organized way to gather information and answer questions.

You must observe carefully in a scientific *inquiry*.

112

## Jupiter

[JOO•pit•er]

The fifth planet from the sun and the largest planet in the solar system.

The Great Red Spot on *Jupiter* is a storm that has been going on for more than 300 years.

111

## investigation

[in•ves•tuh•GAY•shuhn]

A procedure that is carried out to gather data about an object or event.

The student is conducting an *investigation* and recording the results.

113

**kidney**

114

**latitude**

115

**liquid**

116

**local wind**

114

## latitude

[LAT•uh•tood]

An area's distance from the equator.

Lines of *latitude* are parallel to the equator.

113

## kidney

[KID•nee]

The main organ of the execretory system.

You have two *kidneys*.

116

## local wind

[LOH•kuhl WIND]

Wind that results from local changes in temperature.

A land breeze is a *local wind* that blows from the land.

115

## liquid

[LIK•wid]

The state of matter that has a definite volume but no definite shape.

Milk is one kind of *liquid*.

117

## low pressure

COOL
WARM
HOT
H High pressure
L Low pressure
Cold front
Warm front
Temperature (°F)
80s
70s
60s
50s
40s

118

## malleable

119

## Mars

120

## meniscus

118

## malleable

[MAL•ee•uh•buhl]

Easy to shape or to form.

Aluminum is a highly *malleable* metal.

117

## low pressure

[LOH PRESH•er]

An area of warm, less dense air.

Stormy weather comes with a *low pressure* area, shown with a letter "L".

120

## meniscus

[muh•NIS•kuhs]

The curved top of a column of liquid.

You can see the *meniscus* when you measure liquid in a graduated cylinder.

119

## Mars

[MARZ]

The fourth planet from the sun.

*Mars* has the largest canyon in the solar system.

121
Mercury
122
metal

123
metal alloy
124
metallic

## metal

[MET•uhl]

A substance that conducts heat and electricity well and is malleable.

Some structures are made of *metal*.

122

## Mercury

[MER•kyur•ee]

The closest planet to the sun.

*Mercury* is a little bigger than Earth's moon.

121

## metallic

[muh•TAL•ik]

Looking like metal.

*Metallic* objects are often shiny.

124

## metal alloy

[MET•uhl AL•oy]

A solid solution in which a metal or nonmetal is dissolved in a metal.

Bronze is a *metal alloy* of copper and other materials.

123

125

## metalloid

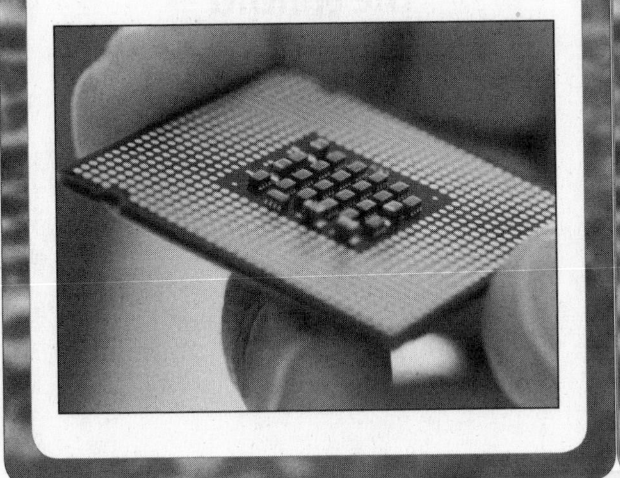

126

## meteor

127

## meteorite

128

## meteorology

126

## meteor

[MEET•ee•er]

A piece of rock that travels through space.

*Meteors* are smaller than asteroids.

125

## metalloid

[MET•uh•loyd]

A substance that has some of the properties of a metal and some of the properties of a nonmetal.

This computer chip is made of silicon, which is a *metalloid*.

128

## meteorology

[meet•ee•uh•RAHL•uh•jee]

The study of weather.

Students use weather stations to help them learn about *meteorology*.

127

## meteorite

[MEET•ee•er•yt]

A meteor that reaches Earth's surface.

Large *meteorites* can leave craters in Earth's surface.

129

## methane

130

## microscope

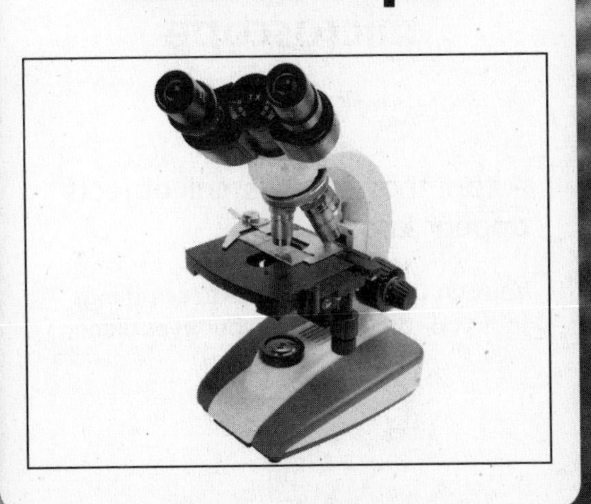

131

## mixture

132

## molecule

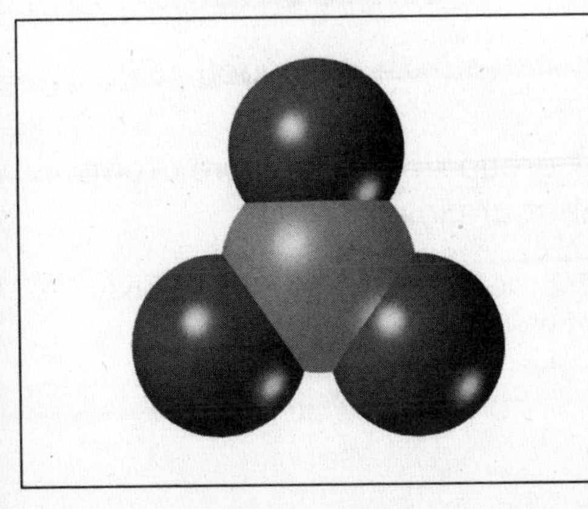

130

## microscope

[MY•kruh•skohp]

A tool that makes small objects appear larger.

You can use a *microscope* to see things that you can't see with your eyes alone.

129

## methane

[METH•ayn]

Natural gas containing carbon and hydrogen.

Gas stoves use *methane* to cook food.

132

## molecule

[MAHL•ih•kyool]

A group of two or more atoms that are joined.

This *molecule* has two different kinds of atoms.

131

## mixture

[MIKS•cher]

A combination of two or more different substances.

Fruit salad is a *mixture*.

133
monsoon

134
moon

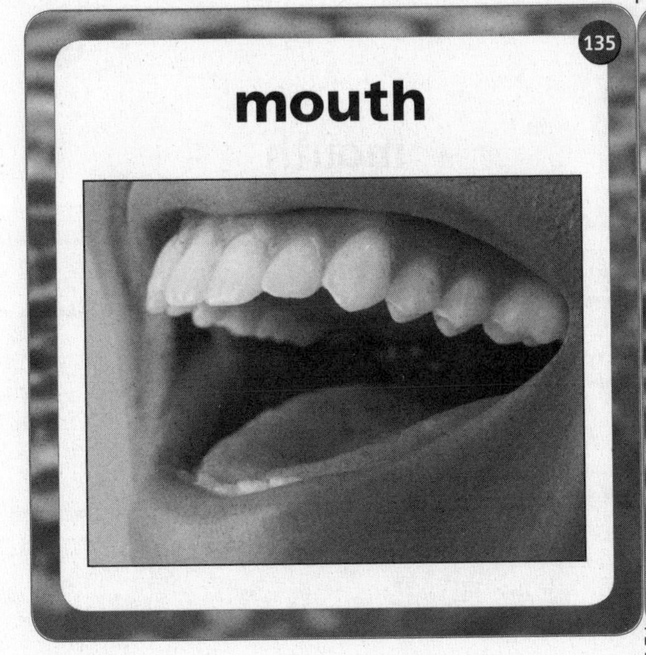
135
mouth

136
multicellular organism

134

## moon

[MOON]

Any natural body that revolves around a planet.

Earth's *moon* causes ocean tides on Earth.

133

## monsoon

[mahn•SOON]

A large wind system that reverses direction seasonally.

*Monsoons* often bring rain.

136

## multicellular organisms

[mul•tih•SEL•yoo•ler AWR•guh•niz•uhm]

A living thing made up of many cells.

A snake is one kind of *multicellular organism*.

135

## mouth

[MOWTH]

The opening through which an animal takes in food, beginning the process of digestion with certain foods.

Your *mouth* is part of your digestive system.

137

## Neptune

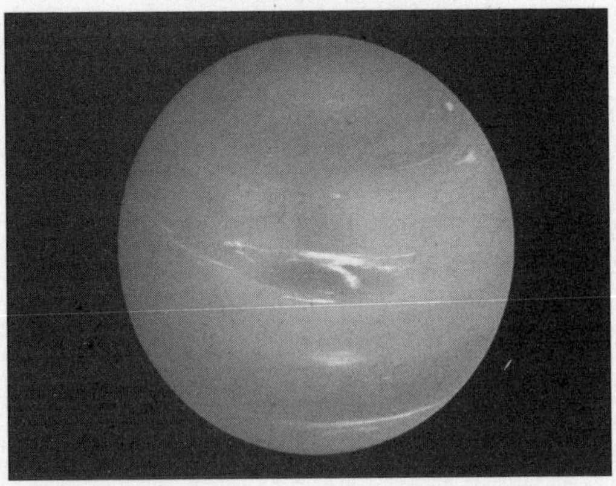

138

## neutralize

139

## neutron

140

## nitrogen

138

## neutralize

[NOO•truh•lyz]

To make chemically neutral, as when an acid and a base are combined.

You can use a base to *neutralize* stomach acid.

137

## Neptune

[NEP•toon]

The eighth planet from the sun.

*Neptune's* blue color is due to methane gas.

140

## nitrogen

[NY•truh•juhn]

A nonmetal element.

*Nitrogen* is a gas that makes up most of Earth's atmosphere.

139

## neutron

[NOO•trahn]

One of the particles in an atom.

*Neutrons* and protons are in the nucleus of an atom.

141

## noble gas

Neon

142

## nonmetal

143

## nonvascular plant

144

## Northern Hemisphere

142

## nonmetal

[nahn•MET•uhl]

A substance that does not conduct electricity and is not malleable.

Coal (carbon) is a *nonmetal*, because it does not conduct electricity and does not bend easily.

141

## noble gas

[NOH•buhl GAS]

An element in the last column of the periodic table that doesn't combine with other elements.

Neon is a *noble gas*.

144

## Northern Hemisphere

[NAWR•thern HEM•ih•sfir]

The half of Earth that is north of the equator, includes North America, Europe, Asia, and parts of Africa.

In the *Northern Hemisphere*, summer is in June, July, August, and September.

143

## nonvascular plant

[NAHN•vas•kyuh•ler PLANT]

A plant without transport tubes to carry water and nutrients thoughout the plant.

*Nonvascular plants* don't grow very tall.

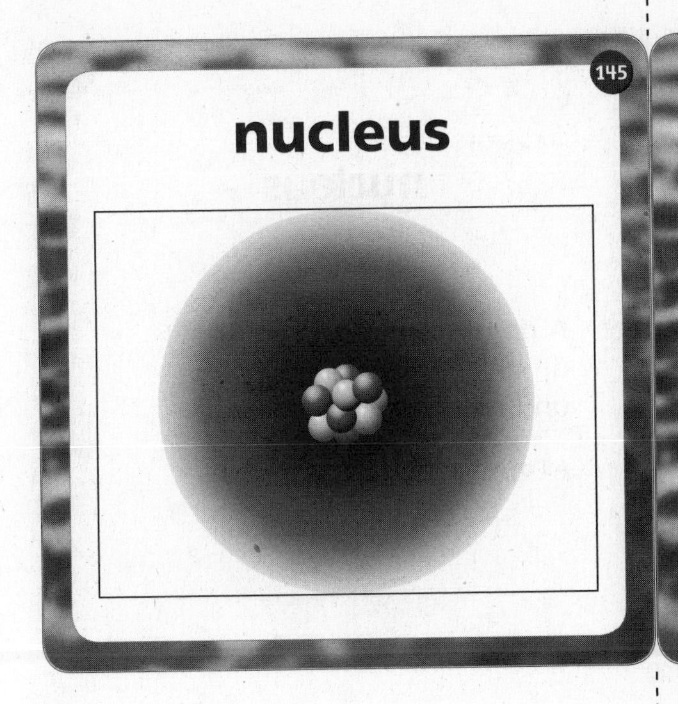
145
nucleus

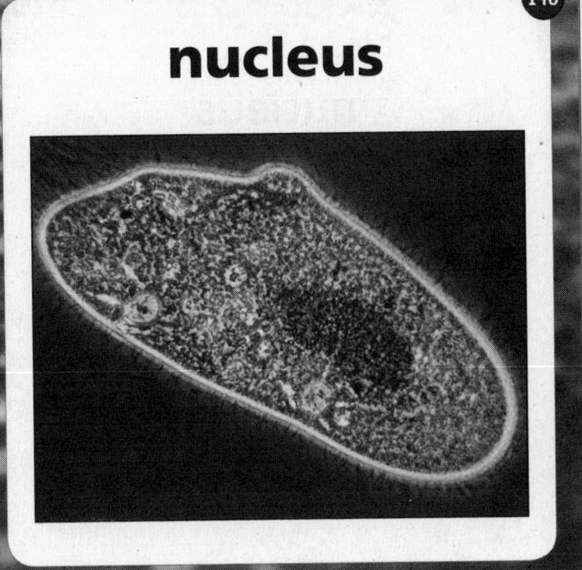
146
nucleus

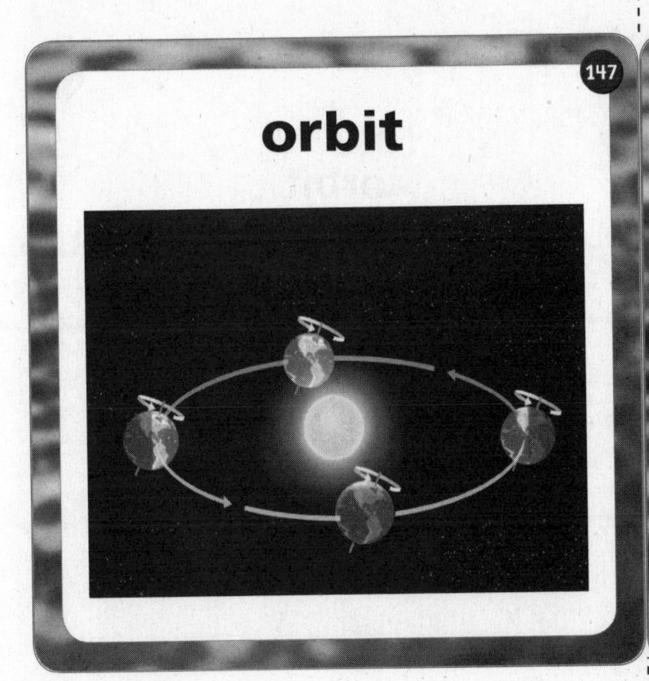
147
orbit

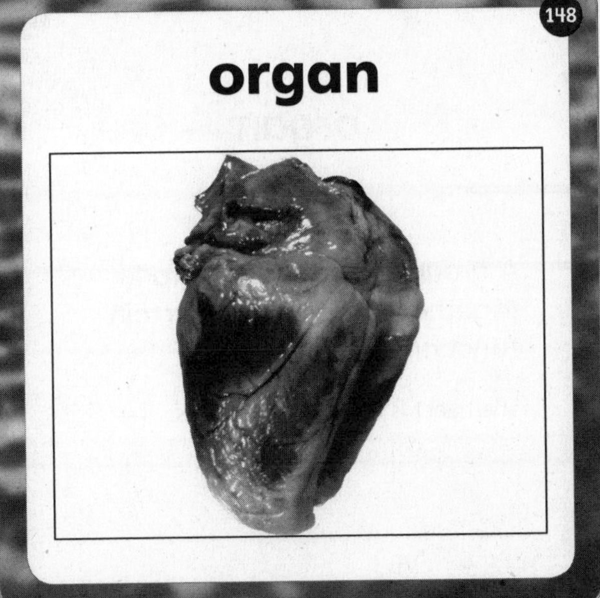
148
organ

146

## nucleus

[NOO•klee•uhs]

In a cell, the organelle that directs all of the cell's activities.

The *nucleus* controls how a cell works.

145

## nucleus

[NOO•klee•uhs]

A dense area in the center of an atom, containing protons and neutrons.

All atoms have a *nucleus*.

148

## organ

[AWR•guh]

A group of tissues that work together to perform a certain function.

The heart is one of your *organs*.

147

## orbit

[AWR•bit]

The path that one body takes in space as it revolves around another body.

It takes 365.25 days for Earth to complete its *orbit* around the sun.

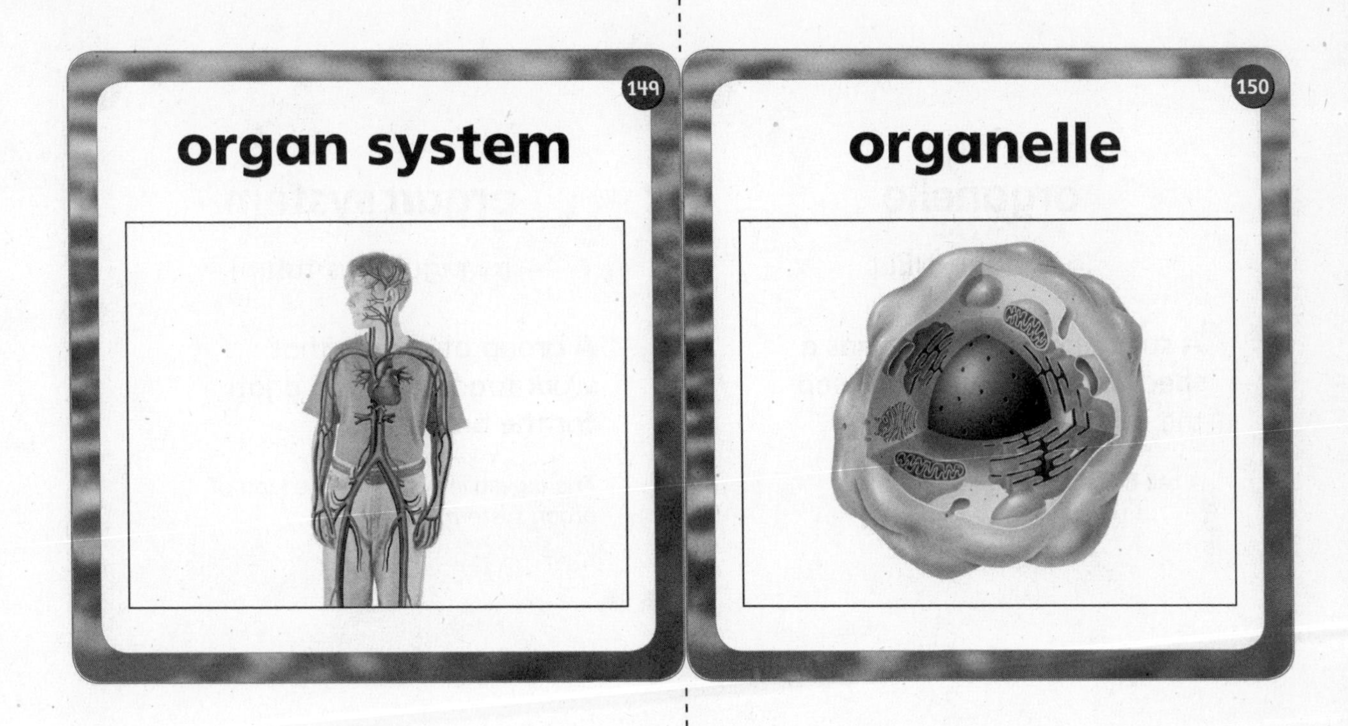
149
organ system
150
organelle

151
organism
152
oxygen

150

## organelle

[AWR•guhn•NEL]

A structure in cells that has a specific function to help keep the cell alive.

A cell has many *organelles*.

149

## organ system

[AWR•guhn SIS•tuhm]

A group of organs that work together to do a job for the body.

The digestive system is one kind of *organ system*.

152

## oxygen

[AHK•sih•juhn]

A nonmetal element that reacts with many other elements.

Humans need *oxygen* gas to survive.

151

## organism

[AWR•guh•niz•uhm]

A living thing.

Plants and animals are *organisms*.

153

**periodic table**

154

**phase**

155

**phloem**

156

**photosynthesis**

154

## phase

[FAYZ]

One of the shapes the moon seems to have as it orbits Earth.

The moon waxes and wanes during its *phases*.

153

## periodic table

[pir•ee•AHD•ik TAY•buhl]

A table that shows the elements arranged by their atomic numbers.

This *periodic table* shows more than 100 elements.

156

## photosynthesis

[foht•oh•SIN•thuh•sis]

The process by which plants make food from carbon dioxide and water and release oxygen into the air.

Plants need light and water to perform *photosynthesis*.

155

## phloem

[FLOH•em]

Vascular tissue that carries food from leaves to the other parts of a plant.

*Phloem* helps transport food throughout a plant.

157

**physical change**

158

**physical property**

159

**physiology**

160

**planet**

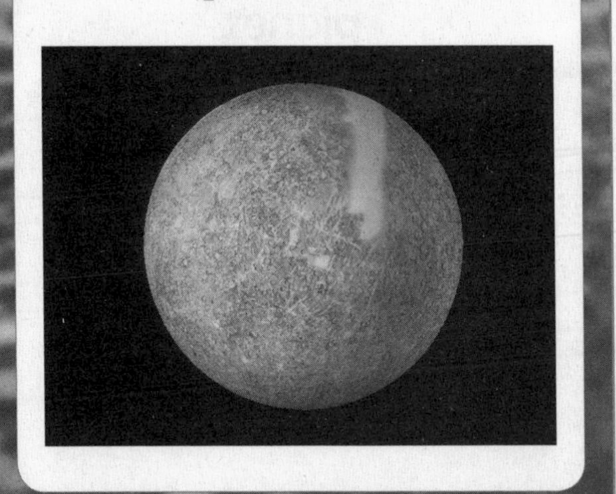

158

## physical property

[FIZ•ih•kuhl PRAHP•er•tee]

A trait–such as color, shape, or hardness–that describes a substance by itself.

Some of the *physical properties* of this animal are its color and shape.

157

## physical change

[FIZ•ih•kuhl CHAYNJ]

A change in which a substance remains the same substance.

Melting is one kind of *physical change*.

160

## planet

[PLAN•it]

A body that revolves around a star.

Our solar system has nine *planets*.

159

## physiology

[fiz•ee•AHL•uh•jee]

The study of how organisms function.

Understanding *physiology* can help runners win races.

161

## planetary

Mercury
Venus
Earth
Mars
Jupiter
Saturn
Uranus
Neptune
Pluto
Sun

162

## Pluto

163

## pollution

164

## precipitation

162

## Pluto

[PLOOT•oh]

The ninth planet from the sun.

*Pluto* is smaller than Earth's moon.

161

## planetary

[PLAN•uh•tair•ee]

Having to do with the planets.

Information about Mercury is one part of *planetary* science.

164

## precipitation

[pree•sip•uh•TAY•shuhn]

Water that falls from clouds to the Earth.

*Precipitation* can be solid, like snow, or liquid, like rain.

163

## pollution

[puh•LOOSH•uhn]

Any change to a resource that makes the resource unhealthy to use.

Factory smoke is a source of air *pollution*.

165

## prevailing westerlies

166

## prevailing wind

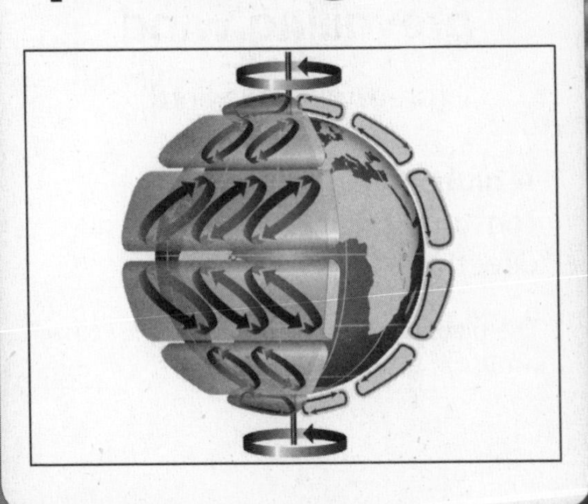

167

## product

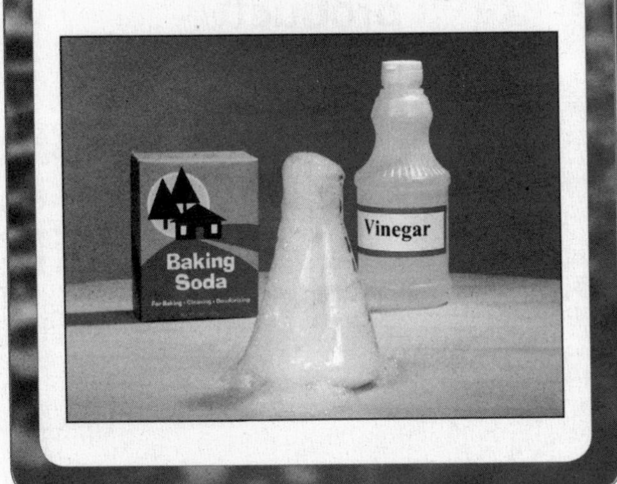


168

## propane

166

## prevailing wind

[pree•VAYL•ing wind]

A global wind that blows constantly from the same direction.

The picture shows Earth's *prevailing winds*.

165

## prevailing westerlies

[pree•VAYL•ing WES•ter•leez]

The prevailing winds over the United States that blow from the west.

The *prevailing westerlies* move weather systems from west to east.

168

## propane

[PROH•payn]

Gas used in heating some homes and for cooking in outdoor barbeques.

This tank holds *propane* for the grill.

167

## product

[PRAHD•uhkt]

A substance that is formed by a chemical reaction.

Carbon dioxide is one of the *products* of combining baking soda and vinegar.

169
properties
170
proton

171
qualitative observation
172
quantitative observation

170

## proton

[PROH•tahn]

One of the particles in an atom.

*Protons* have a positive charge.

169

## properties

[PRAHP•er•teez]

The physical and chemical qualities of an substance, such as size, texture or ability to react with other substances.

One of the *properties* of this glass jar is that it is smooth.

172

## quantitative observation

[KWAHNT•uh•tayt•iv ahb•zer•VAY•shuhn]

An observation that involves numbers or measurements.

The student is using a balance to make a *quantitative observation*.

171

## qualitative observation

[KWAWL•uh•tayt•iv ahb•zer•VAY•shuhn]

An observation that does not involve measurements or numbers.

A description of a smell is a *qualitative observation*.

173

## quarter moon

174

## rain

175

## reactant

176

## reclamation

174

## rain

[RAYN]

Precipitation that is liquid water.

Some forest areas receive a lot of *rain.*

173

## quarter moon

[KWAWRT•er MOON]

The moon phase half way between a new and a full moon.

A *quarter moon* looks like a semicircle.

176

## reclamation

[rek•luh•MAY•shuhn]

The recycling of sewage water.

Sewage water can be used again if it goes through *reclamation.*

175

## reactant

[ree•AK•tuhnt]

A substance that changes during a chemical reaction.

Vinegar and baking soda are *reactants* when you combine them.

177

## recycle

178

## recycling

179

## reservoir

180

## respiration

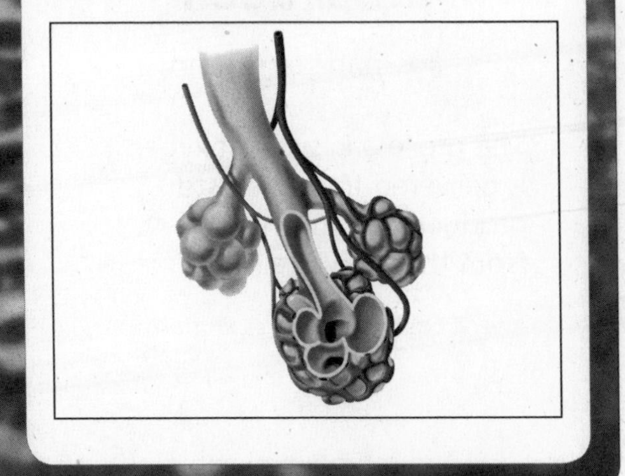

178

## recycling

[ree•SYK•ling]

Making new products from old ones.

*Recycling* is one way to help the environment.

177

## recycle

[ree•SY•kuhl]

To use something again for a new purpose.

Trees *recycle* oxygen from carbon dioxide.

180

## respiration

[res•puh•ray•shuhn]

The process in which oxygen is delivered to and waste products are taken away from the body's cells.

Alveoli are involved in *respiration.*

179

## reservoir

[REZ•er•vwar]

A body of water stored for future use.

*Reservoirs* provide fresh water to many cities.

181

**respiratory system**

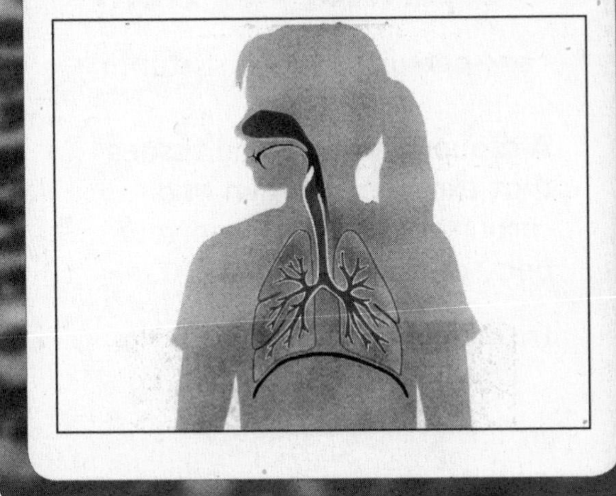

182

**saliva**

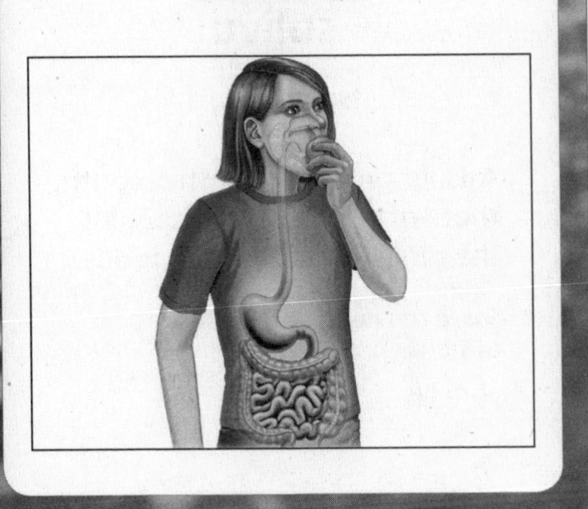

183

**salt**

184

**satellite**

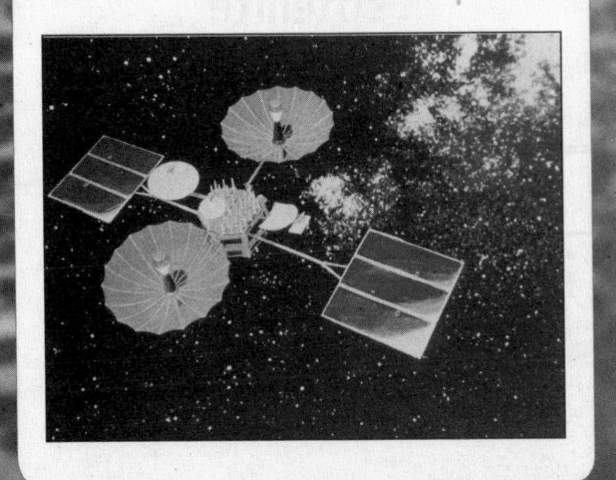

182

## saliva

[suh•LY•vuh]

A fluid produced by the mouth, that softens food, and begins the digestion of certain foods.

*Saliva* can help start digestion of starches.

181

## respiratory system

[RES•per•uh•tawr•ee SIS•tuhm]

A group of organs and tissues that exchange oxygen and carbon dioxide between your body and the environment.

The *respiratory system* includes lungs.

184

## satellite

[SAT•uh•lyt]

A body in space that orbits a larger body.

Some *satellites* are natural, and some are artificial.

183

## salt

[SAWLT]

A substance that is made by combining an acid and a base.

The table salt you use on your food is just one kind of *salt*.

185

## Saturn

186

## scientific method

187

## sea level

188

## seasonal

186

## scientific method

[sy•uhn•TIF•ik METH•uhd]

A series of steps used to plan and carry out an experiment.

You can use the *scientific method* to answer your science questions.

185

## Saturn

[SAT•ern]

The sixth planet from the sun.

*Saturn* has dozens of moons and many rings.

188

## seasonal

[SEE•zuhn•uhl]

Dependent on or determined by the time of year.

Some trees show *seasonal* changes in the colors of their leaves.

187

## sea level

[SEE LEV•uh]

The level of the surface of the ocean, used as a standard in measuring heights and depths.

Some areas of California are below *sea level*.

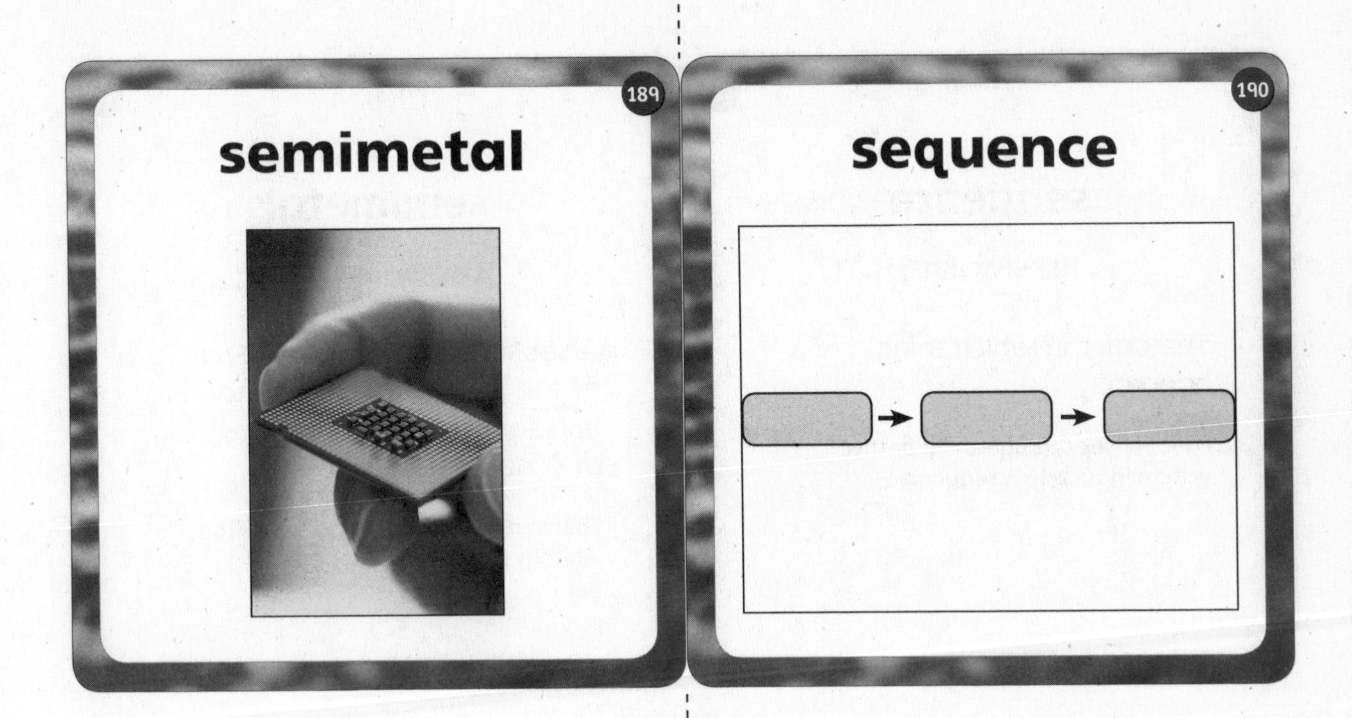
189
semimetal
190
sequence

191
severe weather
192
sleet

190

## sequence

[SEE•kwuhns]

The order in which things happen.

Food moves through the digestive system in a certain *sequence*.

189

## semimetal

[SEM•ee•met•uhl]

A substance that has some of the properties of a metal and some of the properties of a nonmetal.

This computer chip is made of silicon, which is a *semimetal*.

192

## sleet

[SLEET]

Precipitation formed when rain passes through air that is cold enough to freeze water.

*Sleet* makes sidewalks very slippery.

191

## severe weather

[suh•VIR WETH•er]

Extremely bad or dangerous weather, such as hurricanes, tornadoes, or thunderstorms.

It can be dangerous to be outside in *severe weather*.

193
snow
194
solar system
Mercury
Venus
Earth
Jupiter
Saturn
Uranus
Mars
Neptune
Pluto
Sun

195
solid
196
solubility

194

## solar system

[SOH•ler SIS•tuhm]

A star and all the planets and other objects that revolve around it.

The largest planet in our *solar system* is Jupiter.

193

## snow

[SNOH]

Precipitation that is formed when water vapor turns directly into ice crystals.

Flakes of *snow* are ice crystals.

196

## solubility

[sahl•yu•BIL•uh•tee]

The ability to dissolve.

The drink mix has high *solubility*.

195

## solid

[SAHL•id]

The state of matter that has a definite shape and a definite volume.

Sand is one example of a *solid*.

197

# Southern Hemisphere

198

# sphere

199

# star

200

# stomach

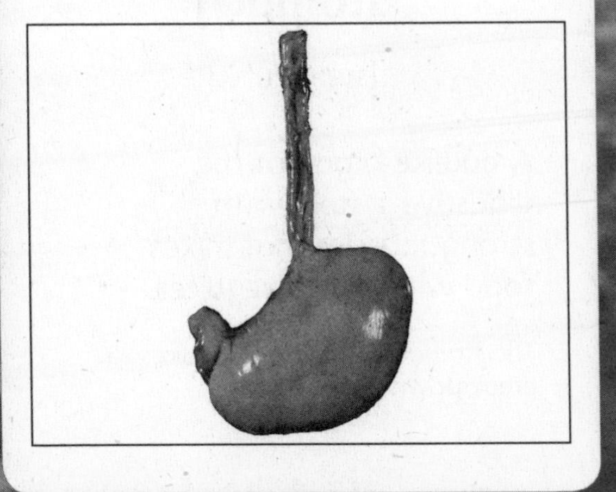

198

## sphere

[SFIR]

The shape of planets and stars.

A *sphere* is the shape of a ball.

197

## Southern Hemisphere

[SUHTH•ern HEM•ih•sfir]

The half of Earth south of the equator, includes South America, Australia, and much of Africa.

In the *Southern Hemisphere*, winter starts in June.

200

## stomach

[STUHM•uhk]

A baglike organ of the digestive system with strong muscles that mixes food with digestive juices.

The *stomach* uses digestive juices to break down food.

199

## star

[STAR]

A huge ball of very hot gases in space.

*Stars* look small from Earth's surface.

201

**stomata**

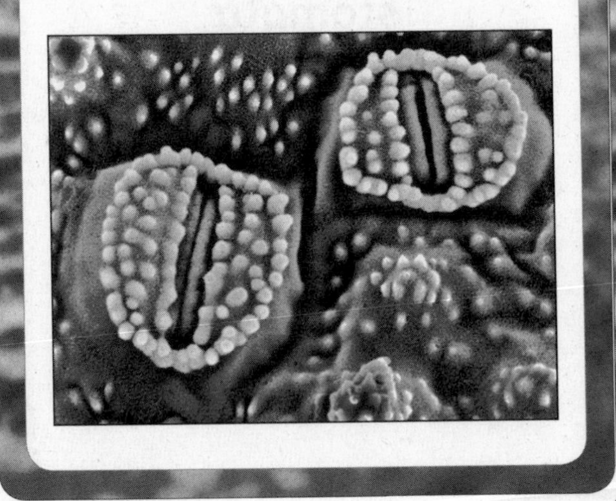

202

**stratosphere**

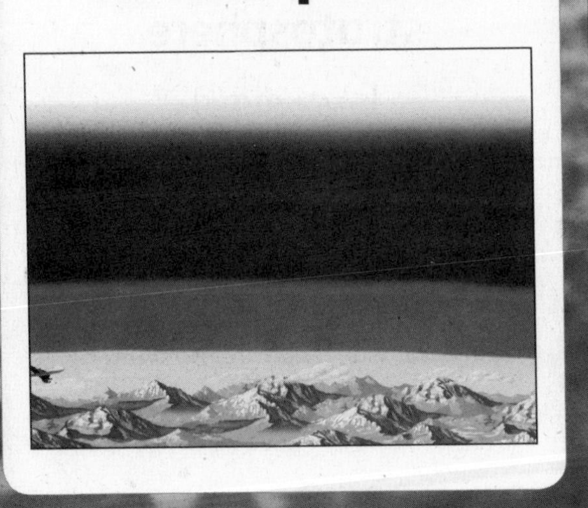

203

**stratus**

204

**structure**

## stratosphere

[STRAT•uh•sfir]

The layer of the atmoshpere containing ozone, and is above the troposphere.

Space shuttles must pass through the *stratosphere* on their way into orbit.

202

## stomata

[STOH•muh•tuh]

On the underside of leaves, tiny holes that release waste products.

Some *stomata* can be seen only with a microscope.

201

## structure

[STRUHK•cher]

In an organism, a part that can be recognized by its shape and other properties.

This bird's wing is a *structure*.

204

## stratus

[STRAT•uhs]

A type of cloud that forms low in the atmosphere, possibly leading to light precipitation.

*Stratus* clouds may look as if they are closer to Earth's surface than other kinds of clouds.

203

205
sublimation
206
sun

207
teeth
208
telescope

206

## sun

[SUHN]

The star at the center of the solar system.

The *sun* provides light and heat energy to our solar system.

205

## sublimation

[suhb•luh•MAY•shuhn]

A change from a solid to a gas without becoming a liquid.

These ice crystals formed through *sublimation.*

208

## telescope

[TEL•uh•skohp]

A tool that scientists use to study parts of the universe not visible to the human eye.

Powerful *telescopes* on Earth and in space reveal faraway parts of the universe.

207

## teeth

[TEETH]

The hard structures in the mouth that grind food into smaller pieces.

People have two sets of *teeth* during their lives.

209

**temperature**

210

**thermal conductivity**

211

**thunderstorm**

212

**tissue**

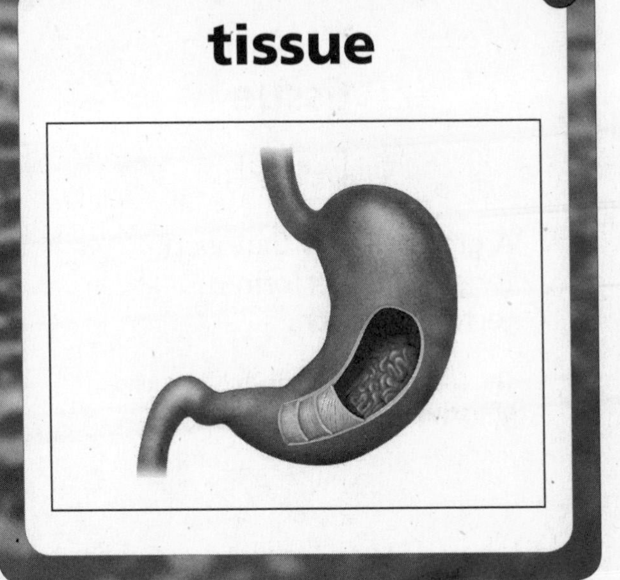

209

## temperature

[TEM•per•uh•cher]

The measure of the quantity of heat in the atmosphere.

Thermometers are used to measure *temperature.*

210

## thermal conductivity

[THER•muhl kahn•duhk•TIV•uh•tee]

The ability of a metal to transfer thermal energy easily.

The wires inside the toaster have high *thermal conductivity.*

211

## thunderstorm

[THUHN•der•stawrm]

A storm with rain, lightning, hail, and thunder.

*Thunderstorms* happen in every part of the country.

212

## tissue

[TISH•oo]

A group of cells that work together to perform a certain function.

The lining of the stomach is one kind of *tissue.*

213

**tornado**

214

**trade wind**

215

**transpiration**

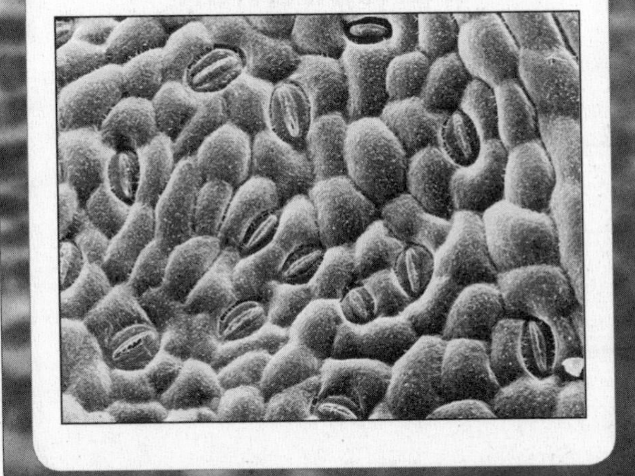

216

**tropical storm**

214

## trade wind

[TRAYD WIND]

A prevailing wind near the equator.

The movement of air near the equator causes *trade winds.*

213

## tornado

[tawr•NAY•doh]

A violently spinning column of air that touches the ground.

The inside of a *tornado* has very low air pressure.

216

## tropical storm

[TRAHP•ih•kuhl STAWRM]

A cyclone in which wind speeds are between 63-118 km/hr (39-73 mph).

*Tropical storms* are less severe than hurricanes.

215

## transpiration

[tran•spuh•RAY•shuhn]

The process by which water moves up and out of plants through tiny holes in their leaves.

Plants recycle water through *transpiration.*

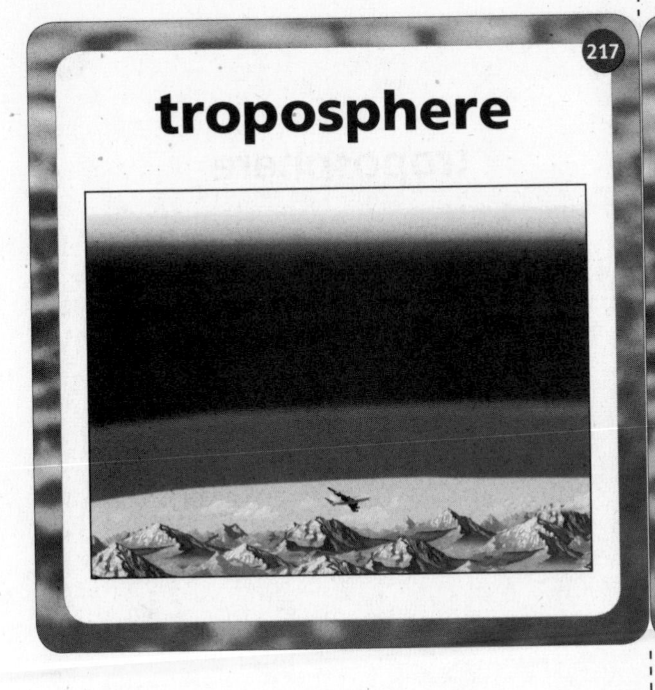
217
troposphere

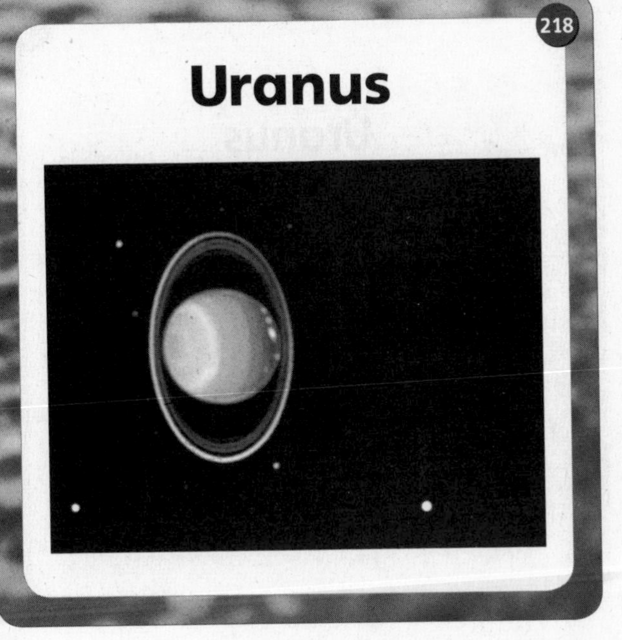
218
Uranus

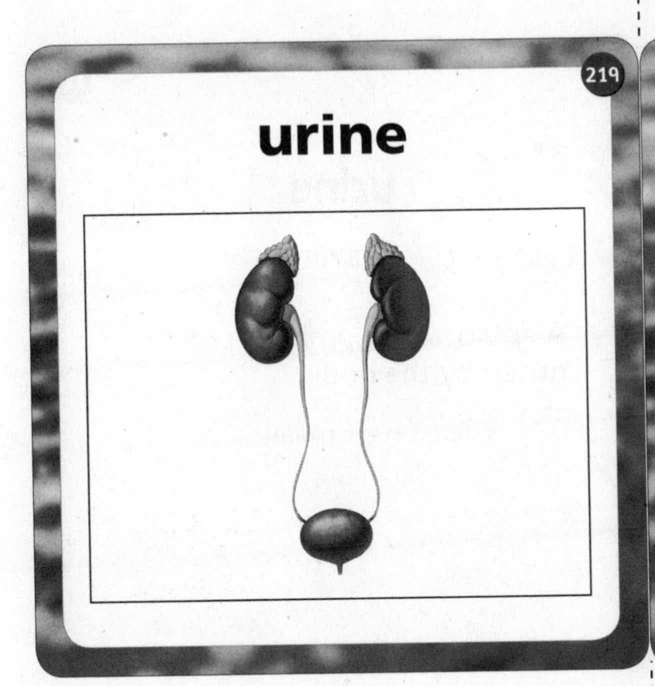
219
urine

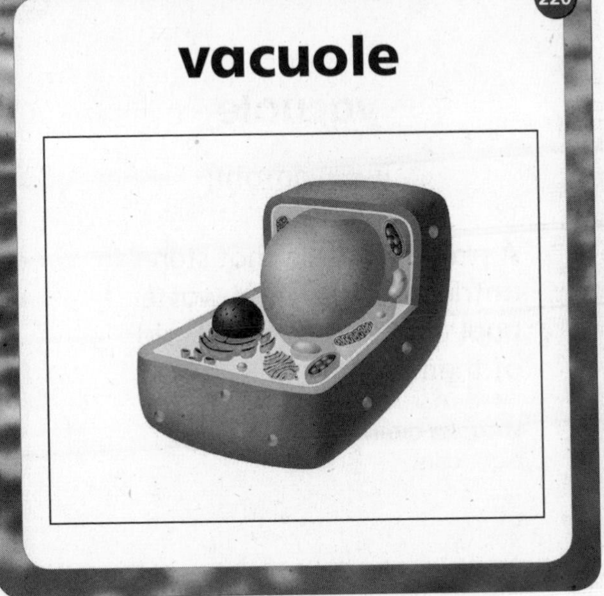
220
vacuole

218

## Uranus

[YUR•uh•nuhs]

The seventh planet in our solar system.

*Uranus* takes 84 Earth years to orbit the sun.

217

## troposphere

[TROH•puh•sfir]

The layer of atmoshpere closest to Earth's surface.

Planes travel in the *troposphere*.

220

## vacuole

[VAK•yoo•ohl]

A plant organelle that stores nutrients, water, and waste until the cell uses or gets rid of them.

*Vacuoles* are like storerooms for plant cells.

219

## urine

[YUR•in]

A liquid waste product produced by the body.

*Urine* is stored in the bladder.

221

## vascular plant

222

## vascular tissue

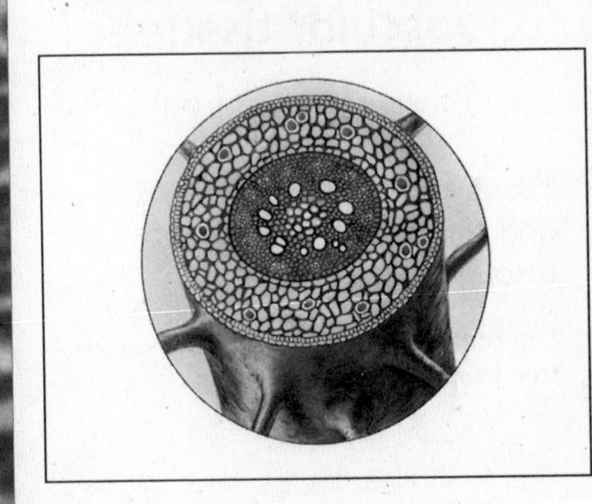

223

## vein

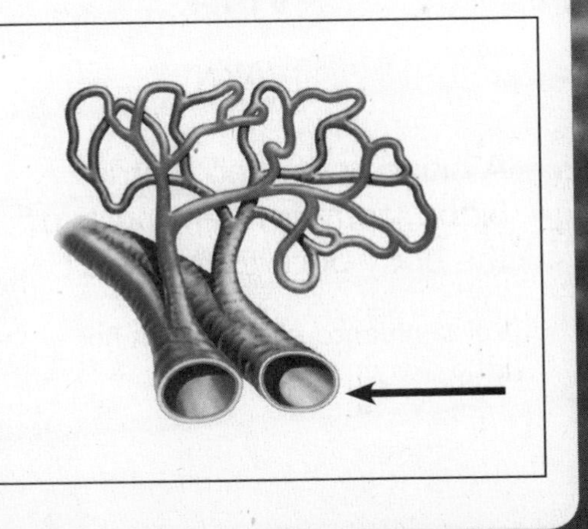

224

## Venus

222

## vascular tissue

[VAS•kyuh•ler TISH•oo]

Tissue that supports a plant and carries water and food throughout the plant.

You can find *vascular tissue* in tree trunks.

221

## vascular plant

[VAS•kyuh•ler PLANT]

A plant with tubes to carry nutrients and water throughout the plant.

Trees are *vascular plants*.

224

## Venus

[VEE•nuhs]

The second planet from the sun.

*Venus* is about the same size as Earth.

223

## vein

[VAYN]

A blood vessel that carries blood from different parts of the body back to the heart.

*Veins* (blue) carry blood that has little oxygen.

225

**vesicle**

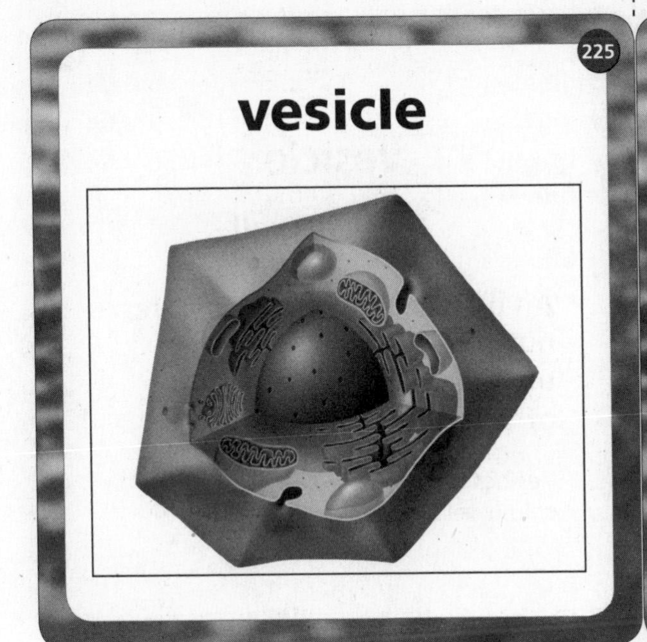

226

**waning**

227

**waste disposal**

228

**water cycle**

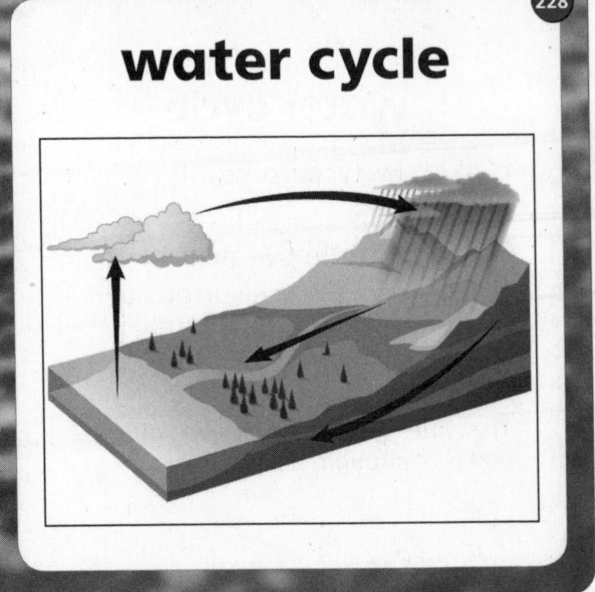

226

## waning

[WAYN•ing]

The process in which the moon goes from full to new.

When the moon is *waning*, the lighted part seen from Earth gets smaller each night.

225

## vesicle

[VEHS•uh•kuhl]

An animal organelle that stores nutrients, water, and waste until the cell uses or gets rid of them.

*Vesicles* are like storerooms for animal cells.

228

## water cycle

[WAWT•er SY•kuhl]

The constant movement of water from Earth's surface to Earth's atmosphere and back to Earth's surface.

The *water cycle* includes evaporation and precipitation.

227

## waste disposal

[WAYST dis•POH•zuhl]

The removal of waste products from an organism.

Every living thing has some kind of *waste disposal*.

229

**water quality**

230

**water table**

231

**water vapor**

232

**watershed**

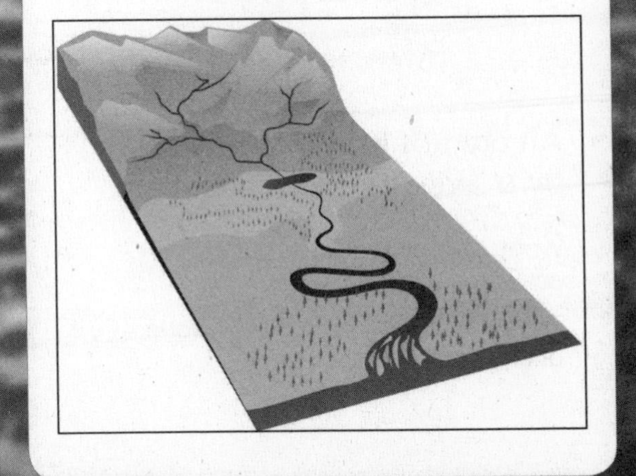

230

## water table

[WAWT•er TAY•buhl]

The top of groundwater.

Surface pollution can get down to the *water table* quickly.

229

## water quality

[WAWT•er KWAWL•uh•tee]

The measure of how safe water is for human use.

The *water quality* in the city is tested regularly.

232

## watershed

[WAWT•er•shed]

An area of land that is drained by a series of creeks and rivers.

Water quality is affected by all the pollution in a *watershed*.

231

## water vapor

[WAWT•er VAY•per]

The gas form of water.

*Water vapor* is formed when water boils.

233

**waxing**

234

**weather**

235

**weather front**

236

**wind**

234

## weather

[WETH•er]

The condition of the atmosphere at a certain place and time.

*Weather* can affect people's activities, especially if it is severe.

233

## waxing

[WAKS•ing]

The process in which the moon goes from new to full.

When the moon is *waxing*, the lighted part seen from Earth gets bigger each night.

236

## wind

[WIND]

The horizontal movement of air.

The *wind* is strong enough to bend these young trees.

235

## weather front

[WETH•er FRUHNT]

A place where two air masses meet.

*Weather fronts* often change the weather.

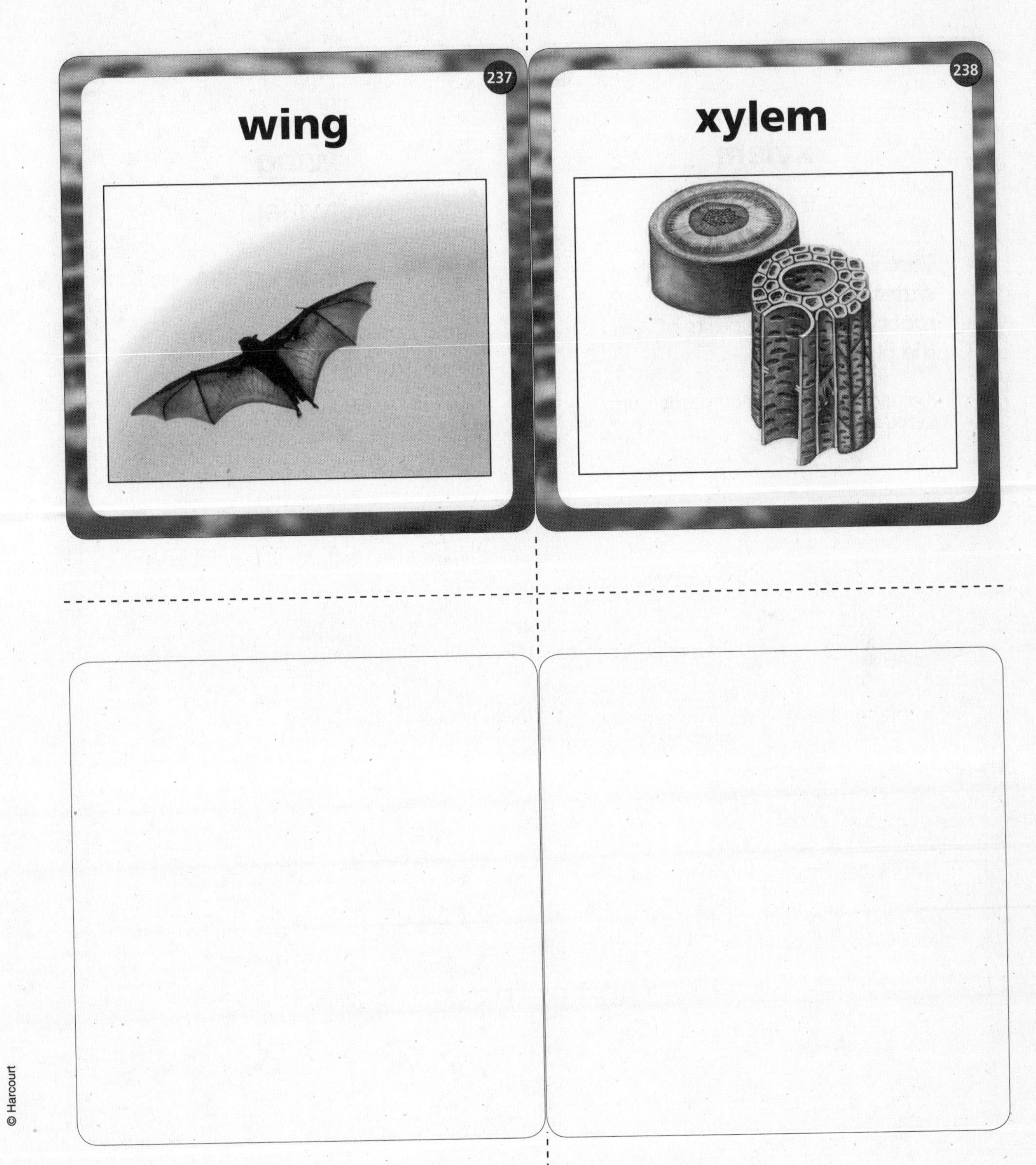
237
wing
238
xylem

238

## xylem

[ZY•luhm]

Vascular tissue that carries water and nutrients from roots to the other parts of the plant.

*Xylem* moves water from the roots up to the leaves.

237

## wing

[WING]

One of the limbs that are typically used for flying by many animals, such as birds, bats, or insects.

A bat uses its *wings* to fly at night.